Innovation by Design

INNOVATION BY DESIGN

Improving Learning Networks in Coastal Management

THE
HEINZ
CENTER

THE H. JOHN HEINZ III CENTER FOR
SCIENCE, ECONOMICS AND THE ENVIRONMENT

The H. John Heinz III Center for Science, Economics and the Environment

Established in December 1995 in honor of Senator John Heinz, The Heinz Center is a nonprofit institution dedicated to improving the scientific and economic foundation for environmental policy through multisectoral collaboration. Focusing on issues that are likely to confront policymakers within two to five years, the Center creates and fosters collaboration among industry, environmental organizations, academia, and government in each of its program areas and projects. The membership of the Center's Board of Trustees, its steering committees, and all its committees and working groups reflects its guiding philosophy: that all relevant parties must be involved if the complex issues surrounding environmental policymaking are to be resolved. The Center's mission is to identify emerging environmental issues, conduct related scientific research and economic analyses, and create and disseminate nonpartisan policy options for solving environmental problems.

About the Sharing Coastal Zone Management Innovations Study

The Heinz Center's Sharing Coastal Zone Management Innovations study was conducted under the terms of a joint project agreement between The Heinz Center and the National Oceanic and Atmospheric Administration's Office of Coastal Resource Management. This report does not necessarily reflect the policies or views of the organizations or agencies that employ the panel members and study sponsors.

Library of Congress Control Number: 200410419 4

International Standard Book Number: 0-9717592-5 1

08 07 06 05 04   5 4 3 2 1

Printed in the United States of America

Additional single copies of this report may be obtained free of charge from

The Heinz Center
1001 Pennsylvania Avenue, N.W., Suite 735 South, Washington, D.C. 20004
Telephone (202) 737-6307   Fax (202) 737-6410   e-mail info@heinzctr.org

This report is also available in full at www.heinzctr.org

Cover illustration: "Salt Water Branches," batik on silk. Copyright © 1991 by Mary Edna Fraser.

CONTENTS

PREFACE

THE CONCEPT and practice of coastal zone management emerged a scant four decades ago from serious soul-searching about how to tackle an array of interconnected problems associated with unprecedented growth and development of our coasts, and with that growth, a host of environmental problems. Numerous studies and popular books in the 1950s and 1960s documented the decline of coasts and the environment generally. One study—the landmark 1969 Stratton Commission report—pointed the way forward, recommending the establishment of a comprehensive national program to manage our coasts. Although the states led the way with coastal management legislation in the 1960s, it was the passage of the U.S. Coastal Zone Management Act (CZMA) in 1972 that provided the fundamental structure and laid out the challenges at the coast.

This study, more than three decades later, is an acknowledgment that we still have a long way to go in meeting this challenge. Further, development within the sciences, generally, and the rapid emergence of a whole new field of information sciences over the last two decades, present coastal managers with many new challenges associated with using new knowledge and application methods to advance their practice in new and creative ways.

It was in this spirit that, in 2002, the National Oceanic and Atmospheric Administration (NOAA) Office of Ocean and Coastal Resources Management—the federal agency responsible for supporting the network of U.S. state coastal management programs developed under the CZMA—asked The Heinz Center to conduct a study of how information sharing in coastal zone management might be improved. The

Heinz Center organized a committee drawn from academia, government, industry, and environmental groups to explore how innovative ideas and practices for coastal problem solving are shared with others who have similar needs, and what improvements are needed.

The committee received a great deal of assistance in conducting this study. We would like to take this opportunity to thank our principal sponsor, the NOAA Office of Ocean and Coastal Resource Management, and the director of its National Policy and Evaluation Division, Ralph Cantral, who also served as a committee member. Thanks are due as well to Margaret Davidson, who helped conceive the study when she was serving as acting administrator of NOAA's National Ocean Service. We also greatly appreciate the assistance of the staff of The Heinz Center, particularly Sheila David, project director for the study, and Judy Goss, research assistant. Sheila refined the committee's scope of work and demonstrated an uncanny knack for drawing the best effort from a group, helping us to be productive and focused throughout the study. Judy provided excellent logistic and substantive support and communication throughout the project. Thanks to you both.

The committee expresses appreciation to those coastal managers who educated us about how they get, use, and share information with their colleagues—especially our interviewees and workshop participants. We really value your time and contributions to the project. Among the workshop participants, we especially appreciate the stimulating presentations of keynoters Robert Kay, coastalmanagement.com Web guru and the brains behind the new United Nations sanctioned OneCoast project, who joined us from Perth, Australia; and Stephen Olsen, director of the University of Rhode Island Coastal Resources Center. Workshop panelists also deserve special thanks for their discussion-stimulating presentations, including Doug Brown, Jeanne Christie, Tracey Crago, Dawn Hamilton, Kalle Matso, Donna McCaskill, Arleen O'Donnell, Miki Schmidt, Suzanne Schwartz, and Carolyn Stem.

All committee members contributed directly to this report and reviewed several drafts. We want to especially acknowledge the heavy lifters who pulled the pieces together, chapter leaders Bob Wayland (Chapter 2), Kem Lowry (Chapter 3), and Madilyn Fletcher (Chapter 4). Finally, we greatly appreciate the fresh perspective our peer reviewers provided near the conclusion of the project: special thanks to Tom Ballou, Jr., Sherwin Alumina Company (retired); Dana Beach, South Carolina Coastal Conservation League; Robert Goodwin, University of Washington Sea Grant;

Nick Shufro, PricewaterhouseCoopers; and Peter Douglas, California Coastal Commission.

This report is aimed at coastal managers in the public, private, and nonprofit sectors at all levels—local, state, regional, and national. However, opportunities for improvements in the networks that facilitate innovation, information sharing, and learning in coastal management require the special attention of its leaders. It is the leaders who can do the most to create a culture and climate that fosters innovation, creative problem solving, and risk taking. The problems and opportunities we face at the coast—thoroughly described in recent reports from the Pew Oceans Commission and the U.S. Commission on Ocean Policy—are looking for that leadership.

JAMES GOOD
Chair

SUMMARY

HOW DO COASTAL MANAGERS—the diverse community of organizations and individuals who make or are otherwise involved in decisions that affect coastal lands, waters, and natural resources—learn about and apply innovative and successful processes, practices, and tools for coastal problem solving? And how can the governmental and nongovernmental organizations that support them be more responsive in providing that information in useful forms and assisting in its adaptation to local situations and needs? Examining these questions was the principal purpose of this study and is the subject of this report.

The impetus for the study came from the U.S. Office of Ocean and Coastal Resource Management (OCRM), the National Oceanic and Atmospheric Administration (NOAA) office responsible for administration of the Coastal Zone Management Act (CZMA). OCRM provides national leadership for thirty-five coastal states and territories eligible to participate in the national coastal zone management program—thirty-four of those states have federally approved programs covering more than 99 percent of U.S. coastlines. OCRM provides policy guidance, grants, and technical assistance to states, and evaluates their performance. With the primary goal of increasing their capacity for providing needed technical assistance, in 2002 OCRM asked The Heinz Center to undertake a study of how innovative ideas and practices are shared within the coastal management community, and to recommend improvements that would foster more effective and efficient information transfer. Recognizing that there are many other organizations and programs involved in coastal management, ORCM asked The Heinz Center to look broadly at technical assistance to coastal managers, not only at the programs that they administer. The audience for the study thus includes local, state, and federal agency decision makers and professionals charged with implementation of the variety of laws and

1

programs designed to address the whole range of coastal conservation and development issues, as well as coastal managers in the private sector involved in these processes.

The day-to-day work of coastal managers is primarily focused on carrying out policies designed to achieve programmatic goals—protecting sensitive habitats and species, reducing the vulnerability of people and property to natural hazards, providing for the recreational needs of diverse visitor populations, and promoting appropriate and sustainable development. To do this well, coastal managers need ready and efficient access to relevant data, information, tools, and processes best suited to each task. Often this entails learning from others who have had to deal with similar issues, and adapting that experience to their own situation. Yet, as OCRM's experience and survey data from the NOAA Coastal Services Center suggest, coastal managers find the process of learning from others cumbersome, hit-or-miss, and too consumptive of time, energy, and resources. How might the learning and technical assistance process be improved? To answer this and related questions, The Heinz Center organized a Committee on Sharing Coastal Zone Management Innovations, drawing on the expertise of representatives from government, academia, industry, and environmental organizations.

Chapter 1 of this report provides background on the problem and details the committee's approach and methods. The committee identified three objectives: (1) to define the problem more clearly by documenting how we presently share coastal problem-solving ideas and practices in government, academic, business, and not-for-profit sectors; (2) to evaluate the strengths, limitations, and outlook for present information-sharing methods and efforts; and (3) to identify ways to improve information sharing and learning, drawing on the experiences of those within and outside the coastal management community. The committee collected data through structured interviews with coastal managers; a coastal manager workshop, designed to explore in more depth questions raised in the survey; and a review of the literature and the experiences of similar communities of practice. The committee also identified examples of innovative practices and information-sharing techniques that were being used within the coastal management community.

Chapter 2 provides an overview and examples of how coastal managers—broadly defined here to include governmental, nongovernmental, and private sector decision makers and professionals—access the data, information, and tools they need. It is very much a demand-driven, decentralized process, with constraints and limitations imposed by available knowledge, skills, data accessibility, resources, and training.

Chapter 3 examines the important roles that networks play in information sharing and learning in coastal management. This focus is in part based on the clear message the committee received from its survey and workshop participants that networking was central to their learning about new ways to solve problems. The committee examined various types of networks for their potential to serve the information-sharing and learning needs of coastal managers.

Chapter 4 examines the rapid advances in communication, information, and sensor technologies and how they have transformed the ways in which coastal managers learn about and share innovations. Although these new technologies are enormously powerful or promising, they require significant investments of time, financial resources, staff, and energy. Capacity and access to technology varies. It is difficult to determine which technologies will be good investments at any point and what kinds of organization, training, and human connections will be needed to put them into practice. The challenge, then, is to make technology serve coastal managers and decision makers, not the other way around.

The findings summarized below reveal how coastal managers are gathering and evaluating information today, how networks are fostering innovation and information sharing, and how technology is contributing to the process. Specific recommendations then follow about how to foster innovation and improve information sharing in coastal management.

HOW COASTAL MANAGERS LEARN TODAY

Coastal management today has its roots in an array of environmental policies initiated more than thirty years ago—the Coastal Zone Management Act and the Clean Water Act are prime examples—as well as the rise of citizen activism during the same era and the private sector response to these new policies. Coastal policies, the citizen activism that sustain them, and business response have evolved over the years. Emphases today are on sustainability, ecosystem-based and precautionary approaches, integrated management, and transparent decision making, among other themes. Other factors bearing on the evolution of coastal management are rapid advances in information technologies and technical tools, improved scientific understanding of complex coastal and ocean problems, increased

sophistication of planning and coordination processes, and more focused
outreach and technical assistance programs.

When coastal managers need to learn about innovative practices
and problem solving elsewhere, they cast a broad net. New and evolving
technologies like the Internet, the World Wide Web, and powerful search
engines play an increasingly important role in this process. Nonetheless,
tapping into personal and organizational networks is even more important.
Coastal managers are busy people, so the information search process is
largely demand driven. They seek out those they know and trust first—
people and organizations that are part of their personal networks. Often,
client coastal managers (e.g., state coastal program managers) seek information about innovative approaches from their program sponsors (e.g.,
OCRM). This is logical because the managers must go to these sources for
policy guidance, funding, and other program implementation activities.
Nevertheless, this "stovepipe" model may be inhibiting the flow of innovative ideas and information among programs with similar goals, and may
even set up competitive situations where organizations attach themselves
to or become identified with particular innovations, much the same
as private sector companies protect their proprietary interests in ideas.
Coastal managers also seek information from other coastal management
programs or from organizations whose primary mission is to provide data,
information, technical assistance, and outreach services—examples include
university-based Sea Grant extension programs, the NOAA Coastal Services Center, and more recently, the coastal training network of the
National Estuarine Research Reserves.

What drives the diffusion and adaptation of good ideas in coastal
management? New public policy initiatives are often drivers for both
innovation and diffusion. An example is the National Estuary Program,
established through 1987 amendments to the Clean Water Act, which
helped catalyze the "watershed approach" to estuarine and aquatic resource
management. Another example is NOAA's Coastal Zone Enhancement
Grants Program, part of 1990 amendments to the CZMA. That program
provided support to states to experiment with new approaches to address
chronic coastal problems such as cumulative impacts of development,
increased demand for public access, and habitat protection. The result was
a new generation of more sophisticated, problem-focused coastal management programs.

Another important factor in the diffusion of coastal management
innovations has been the role of "champions"—individuals whose long-

term commitment to developing, describing, publicizing, marketing, and providing technical assistance for innovative coastal management practices has made a difference. Conferences such as those of the Coastal Society and the biennial Coastal Zone meetings build and sustain the formal and informal networks and personal relationships that are so important for information sharing, not just during the events, but for years afterward. More focused workshops are important as well for in-depth, specialized learning. The continued seeding of the coastal management profession with recent graduates of specialized marine science and policy graduate programs is another important factor in the diffusion of new ideas—often learned in a more theoretical context, but then put to the test in the field.

Looking ahead, coastal managers envision significant changes in how they learn about and apply new ideas to solve increasingly complex problems, especially in the role that technology might play in the field. They also see the need to maintain and strengthen traditional mechanisms for information transfer—conferences and targeted workshops will continue to be important—but more and better use of new networking technologies, such as video conferencing and Internet streaming, will also be necessary. Coastal managers would like to see a reliable, quick-response "consulting service" available as well, providing technical assistance to adapt and tailor ideas to particular situations. Information purveyor networks—Sea Grant, the Coastal Services Center, the National Estuarine Research Reserves—will remain vital, but will need to be better integrated and coordinated across agencies, organizations, and levels to be most effective. Coastal managers also noted barriers and constraints to optimizing the development and diffusion of innovations in coastal management practice. Major problems that need to be addressed include information overload, the "stovepipe" information flow problem noted earlier, limited resources for travel to conferences and workshops, the risks of experimentation, and the bias against reporting failures (and consequent loss of learning opportunities).

Two closely linked themes pervaded the committee's interactions with coastal managers. One was the importance of people-centered networks in fostering innovation and spreading information. The other was the growing role of technology. As we consider how to develop more conscious, robust learning networks in coastal management, these two themes—one focused on human interactions and the other on computer-based connections—must be viewed as fully interdependent. Neither alone is sufficient.

LEARNING NETWORKS AND
COASTAL MANAGEMENT

Networks are ubiquitous in all collective human endeavors. In coastal management, there are hundreds, perhaps thousands, of networks, often self-organizing and sustaining. Some networks are organized around issues such as natural hazard mitigation or habitat restoration, and are inherently multidisciplinary. Other networks are based on professional identities and training; thus we have disciplinary organizations for planners, engineers, or wetland ecologists. Sectoral networks, based on broad institutional connections, are another type—fishing industry organizations or government agencies concerned with pollution control are examples. Political networks of all varieties operate in coastal management as well, attempting to affect public policy and promote agendas. Political networks overlap considerably with other kinds of networks. Issue networks, for example, are usually strongly linked to particular problems and policy solutions.

Virtually all of these networks have learning as one of their principal functions—thus the term *learning networks*. Some organizations are formed consciously to function as a learning network, developing, testing, and transferring innovations and information—the ten-year-old NOAA Coastal Services Center is such an organization serving coastal managers.

We can distinguish learning networks by their structure, purpose, membership, formality, governance, and other features. Some are quite formal and structured, such as a state coastal management program or a project team within such a program. Professional networks (e.g., the Coastal Society), sectoral networks (e.g., the American Association of Port Authorities), and issue networks (e.g., Sea Grant's hazards network or HAZNET) each have different degrees of formality and structure. Collaborative networks (e.g., the Ocean Governance Study Group) and communities of practice (e.g., the Locally Managed Marine Area Network) may be formal or informal, but are generally characterized by their membership, which is often by invitation only.

A key feature of these and other coastal management learning networks is their ownership by members. Ownership implies participation and involvement in decisions about how the network can be most useful for learning. This does not mean that all networks need to be highly structured and organized. Quite the opposite—networks need to serve their members and thus may take many forms.

Learning networks can be differentiated by a number of other important features: their organizational culture and leadership; the resources they have at their disposal; the roles of members, particularly who will act as the "node" of the network; and network connections, internally and externally. With respect to culture and leadership, a key issue is the degree to which an organization and its leaders choose to be innovative—that is, to empower its members to experiment and take risks, to share ideas and practices, and to learn from others. Being innovative and sharing ideas, of course, is always easier when resources are abundant (which they almost never are). Effectively connecting multiple learning networks to add value to all is another challenge.

Recognizing and transferring useful innovations and information to others, and learning about, acquiring, and adapting or tailoring the innovations of others, are key functions of an effective learning network. But they are not necessarily simple functions. They often require the time and energy of a network's most experienced, savvy members. Learning network members need to be able to understand the importance of context to the successful adaptation of an innovation. They need to be aware of the substitutability of institutions (or not) from one location to another, the resources needed to transfer an innovation, and the relative complexity of both the innovation and the transfer process.

Most coastal management networks probably function fairly well with respect to innovation—learning, adapting, and sharing ideas and information. Few, however, have been seriously evaluated as true learning networks, as defined here. Most coastal management organizations address innovation issues intuitively, in an ad hoc, haphazard manner. Although this approach seems to work at some level, it is intriguing to imagine what a more conscious, deliberate effort to build effective learning networks within the coastal management community might accomplish. The continuing rapid development of information technologies makes such an undertaking feasible. The huge pressures on coastal areas and resources, the resulting problems, and the need for more sustainable development make it necessary.

TECHNOLOGICAL CHANGE AND COASTAL MANAGEMENT

The practice of coastal management is being transformed by remarkable technological developments in advanced communications, information management, and remote sensing. The Internet, the World Wide Web, geographic information systems (GIS), real-time observing systems, and other technologies enhance learning and create new methods for learning. The result is a flood of numbers and types of learning networks available to coastal managers and a deluge of information—some relevant and useful to coastal problem solving, much of it not. Some coastal managers and organizations are open to the new technologies, while others have tried to channel or filter the flow, creating personal and virtual firewalls. The challenge for coastal managers is to incorporate these new technologies into their existing learning networks in ways that enhance, rather than degrade, the learning and information-sharing process. This will require much more attention to the interfaces among people, organizations, and new technologies. Although the direction of technological change is unpredictable, the advent of wireless communications, artificial intelligence, software tools such as bots, and ever-increasing computing power will continue to transform the practice of coastal management. Suitable governance arrangements for incorporating new technology into coastal management, and for collecting, documenting, processing, and applying data and information will be needed.

RECOMMENDATIONS

Coastal managers will continue to learn in many ways—from face-to-face meetings, conferences, workshops, Internet searches of the World Wide Web, and many other sources—and they will also continue to innovate and share what they learn. Nevertheless, the community has many opportunities to improve the learning potential of its myriad networks, opportunities that are expanding daily as technology changes. Barriers and constraints must be removed to take fuller advantage of these opportunities. Incentives to generate and accurately document innovative practices are a clear need, as are standards, so that best practices can be validated as widely applicable and repeatable. More concerted attention to diffusion of innovations is also needed, with provisions for documenting and collecting experience, facilitating the searching process, and adapting and tailoring information to local contexts. The five recommendations presented here address these and other issues identified by the committee (Box S.1).

Box S.1   Recommendations of the Committee on Sharing Coastal Zone Management Innovations

- The committee recommends that organizational leaders evaluate and strengthen coastal management learning networks.
- The committee recommends that NOAA develop and manage a compendium of examples and studies of the best coastal and ocean management practices, supported by a network of experts.
- The committee recommends the expansion of cross-training of personnel to broaden mutual understanding of coastal management problems, practices, and uses of technology.
- The committee recommends that coastal management organizations and agencies increase the use of communication technology for real-time, distributed learning.
- The committee recommends that organizations institutionalize a learning process about interactions between technology and coastal management.

- **The committee recommends that organizational leaders evaluate and strengthen coastal management learning networks.**

The exchange of ideas and information within and among learning networks in coastal management plays a vital role in stimulating innovation and facilitates the transfer of ideas to others who adapt and tailor them to fit their needs. Given daily demands and limited resources, however, little explicit attention is given to nurturing these networks to realize their potential. For innovation and network-based learning to be fixtures in coastal management practice, the individuals that lead organizations need to encourage experimentation, risk taking, and similar behaviors not generally found in the field. This is especially true in the public sector, where political forces sometimes push in the opposite direction. The findings of this report suggest a number of needed actions to improve and strengthen learning networks in coastal management.

- *Increase Support to Learning Networks.* NOAA, as the nation's public sector leader for ocean and coastal affairs, should create an environment within and outside government that supports a wide range of existing and new learning networks designed to encourage and share innovative and best practices for coastal management. Because of the nature of learning networks—informal ones often being more nimble than formal—care should be taken not to provide too much structure. Organizations such as the Office of Ocean and Coastal Resources Management and Coastal Services Center have or can develop the research capacity, multiple linkages, and technical facility to become stronger nodes and facilitators of learning network nodes. Other organizations within and outside government that have established effective learning networks—Sea Grant, the U.S. Environmental Protection Agency (USEPA), and Restore America's Estuaries, for example—should participate in this process as full partners, recognizing their unique potential contributions. All such learning networks should be encouraged to examine their capacity and effectiveness in fostering innovation and learning.

- *Leadership Training for Innovation.* The NOAA Coastal Services Center, in collaboration with OCRM, the USEPA, Sea Grant, and academic programs in marine affairs and policy, should develop and deliver a training program for coastal management leaders that emphasizes the potential of learning networks to promote knowledge-based problem solving. Training about learning networks and their roles in innovation, adaptation, and change needs to be fostered at all levels, including formal education. In particular, graduate students being trained for coastal management should be steeped in learning networks and innovation processes.

- *Collaboration among Information Purveyors.* Organizations for which outreach is a principal function* should work more closely to increase their effectiveness with coastal manager audiences, and reduce duplication and competition. Partnership building methods for Coastal Training Programs at National Estuarine Research Reserves serves as a good model.

- *Multinodal Learning Network.* NOAA should establish a multinodal learning network for identifying, documenting, validating, collecting, searching for, and tailoring best coastal management practices to local contexts. The challenges are to identify practitioner information needs, preferred communication media, and incentives to encourage practitioners to treat the network node as an accepted information broker and network facilitator.

---

* Sea Grant Extension, the Coastal Services Center, and the Coastal Training Program are examples highlighted in Box 2.4, pages 39–42.

■ **The committee recommends that NOAA develop and manage a compendium of examples and studies of the best coastal and
ocean management practices, supported by a network of experts.**

The committee proposes that NOAA establish a compendium of
peer-reviewed case studies and examples of innovative or successful coastal
management practices—best coastal and ocean management practices, or
BCOMPs. The committee acknowledges that this idea is not new and
that it raises concerns about the maintenance, management, and utility of
such a repository. Case studies become quickly outdated, quality is mixed,
adaptation and tailoring is difficult, and best practices may be misapplied.
There are ways to overcome some of these shortcomings:

- Establish the system as a Web-based, searchable portal (similar to
  and perhaps in conjunction with GIS data nodes being established).
- Institutionalize incentives, mechanisms, and standards for submitting BCOMPs.
- Provide sufficient structure to make the system self-organizing,
  logical, and demand-driven.
- Link the compendium to an expert network that can assist
  coastal managers in the adaptation and tailoring of BCOMPs to
  unique local contexts.

Many existing coastal management networks would have roles
to play in such a system. OCRM and its client state coastal programs
could identify candidate BCOMPs through coastal zone management
performance and other evaluation processes. The NOAA Coastal Services Center could serve as a national node or coordinator for such a system. Many of its own technology-based projects currently available on
the World Wide Web are candidate BCOMPs, and Coastal Services
Center staffers could help identify others. Sea Grant's university-based
programs in every coastal state could also serve as nodes in such a system.
Other coastal management agencies and programs—including the USEPA,
the U.S. Fish and Wildlife Service, NOAA Fisheries, the National Park
Service, Coastal America, state agencies, nongovernmental organizations,
and private-sector businesses and trade organizations—might also identify, nominate, or prepare BCOMPs, using specified criteria and formats,
and serve as additional nodes.

A Web-based, online compendium is one possible approach, operated by a consortium of university-based marine affairs and policy graduate programs, faculty, students, and others providing well-researched, documented cases meeting high standards of relevancy and quality. Incentives for submissions might include awards for "most innovative practice" or other recognition. Examples from international coastal management could be included through regional nodes. Specialists, available as consultants to assist in assessing applicability and tailoring lessons, could be a feature of such a system. Use of artificial intelligence agents (bots) to guide the searcher to appropriate information or case examples in the compendium could be explored. Organization of such a system of BCOMPs will be a challenge—a traditional structure (shown in Table S.1) is one alternative, but a more open system based on key words or phrases might be another.

Finally, standard formats for presenting case examples and studies will be needed to capture, at the very least, the what, who, why, and how, along with outcomes and lessons learned. This last category also opens up the door for examples of what did not work and what adaptive learning took place in the process.

■ **The committee recommends the expansion of cross-training of personnel to broaden mutual understanding of coastal management problems, practices, and uses of technology.**

Training across sectoral, professional, political, governmental, hierarchical, and other boundaries helps coastal management professionals learn, understand other perspectives, and appreciate the constraints and limitations encountered by their colleagues in different organizations. The diversity within the field of coastal management makes this cross-training even more necessary.

The committee recommends that NOAA, the USEPA, and other federal coastal agencies, in collaboration with the Coastal States Organization and nongovernmental organizations, expand the use of the federal Intergovernmental Personnel Act (IPA) in the field of coastal management to include professionals from state and local governments, nongovernmental organizations, academia, industry, and others who might be eligible. A guide to IPA opportunities in coastal management should be developed by NOAA and a program established to coordinate short-term assignments. Transfers should be two-way, ideally switching specific jobs.

Table S.1 Preliminary Organization for a Compendium of Best Coastal and Ocean Management Practices

Category	Examples
Applied research and technology	Defining management-related research needs, e.g., National Estuary Program experience Inventories Rapid assessment techniques Data and information nodes GIS applications and online mapping Remote conferencing
Area planning	Special area management planning (SAMP) process with examples Coastal land use planning Development buffers and setbacks Shoreline and marine zoning Waterfront revitalization National Estuary Program processes
Regulatory measures	Permit programs One-stop systems Advance planning for permitting
Nonregulatory measures	Conservation easements Transferable development rights Financial incentives
Land and water area management	Direct land management, e.g., National Wildlife Refuges, parks and seashores, Estuarine Research Reserves Land trusts Incentive-based programs, e.g., Wetlands Reserve Program
Public and user education for resource management	Interpretive signage and displays Regulatory signage Private sector collaborations, e.g., through the tourism industry University-based outreach
Ecosystem restoration, enhancement, and creation	Watershed restoration Species and habitat restoration Invasive species control Estuary and streams rehabilitation
Citizen engagement	Volunteer training and monitoring Marine and watershed stewards programs Watershed councils and associations
Coordination and collaboration	Inter- and intra-governmental coordination Public–private sector partnerships Area-based collaborations Policy-based tools, e.g., federal consistency Regional organizations

- **The committee recommends that coastal management organizations and agencies increase the use of communication technology for real-time, distributed learning.**

Some real constraints to learning in coastal management are workload demands and limits on travel funds to attend workshops and conferences—highly valued learning venues. The committee recommends increased experimentation with the design and execution of large-scale video conferencing to expand the audience and access to national conferences and workshops on coastal management practice. The committee envisions national events with regional and local counterparts, as well as interactive participation from a variety of sectors.

The committee recommends that national coastal management agencies and organizations—NOAA, the USEPA, the Coastal States Organization, and others—work with their local clients and with nongovernmental organizations and the private sector to develop a five-year plan to advance the use of current and emerging communication technologies in new, modified, or expanded learning networks. The cost of the technology is not a major issue, given available communication systems and software at universities, community colleges, state capitals, and on individuals' desktops at work and at home. Conference or workshop design for effective learning will be the biggest challenge and may require considerable experimentation and evaluation.

- **The committee recommends that organizations institutionalize a learning process about interactions between technology and coastal management.**

Because of the rapid advances in communication, information, and sensor technologies, the committee recommends that the NOAA Coastal Services Center, in collaboration with the Cooperative Institute for Coastal and Estuarine Environmental Technology, OCRM, the USEPA, and Sea Grant, establish and deliver a workshop series on Coastal Management and Emerging Technologies. We envision an event every two or three years that brings together practicing coastal managers (and their information needs and desires), applied coastal management technologists (e.g., Coastal Services Center staff, other agency and organization specialists), and pure technologists. Themes might differ from workshop to workshop, but the overall purposes would be to learn from one another, match needs and desires with potential technological solutions, develop pilot project proposals, and plan diffusion strategies for proven applications.

The Coastal Geotools Workshops, organized periodically by the NOAA Coastal Services Center, may be a prototype for what is envisioned as a more broad-based series. Such a workshop series should be led by a consortium of coastal management agencies at the national or regional level, but should also involve nongovernmental and private-sector organizations and professionals. Participation, particularly among coastal managers, should include leaders and line staff from local, state, regional, and national organizations and networks.

Although coastal managers have often lagged behind other constituencies in taking advantage of new technologies to share knowledge, they still have the potential to revolutionize their use of information to better protect coastal resources. These recommendations are a starting point, the beginning of a dialog, to move coastal management practice toward the goal of being *innovative by design*—where learning through our myriad networks is fully integrated into organizational cultures and individual practice.

1

INTRODUCTION

COASTAL MANAGEMENT as a distinct practice emerged just a few decades ago, when ideas and information were exchanged through mostly conventional means. Scientists and coastal planners gave talks and presented posters at conferences and workshops, as they still do. Field trips and tours organized as part of these events highlighted problems and success stories. Agency experts prepared and distributed reports and guidelines. Academics researched problems and systematically evaluated methods to address them, reporting their results in new periodicals like the *Coastal Zone Management Journal*. Face-to-face meetings, telephone conversations, the U.S. Postal Service, and later the fax machine played key roles in the development of ideas and movement of information to address coastal problems. Working with these communication tools, professionals and concerned citizens alike drew from their personal experience, new state and federal legislative mandates, and a palpable sense of urgency to create a new practice called coastal zone management. At the time, the demand was great for scientific data and information about coastal resources and use, for tools to interpret this information, and for strategies and processes to apply it for problem solving. And the information flowed freely, albeit by slower and less sophisticated means than today.

Now, little more than a decade into the Information Age, the Internet and World Wide Web give us ready access to incredibly diverse data sets and information, at levels and speeds we could not imagine just a few years ago. Technological advances in many fields, including remote sensing and in-situ data collection and processing, computing and software, telecommunications, and data integration and distribution over the Internet, are transforming the practice of coastal management. Geographic information systems (GIS), with their powerful computer-based information handling, storage, analysis, and communication tools, are contributing greatly to this transformation.

But are coastal managers today any better off, information-wise, than they were twenty years ago? The answer, based on information collected for this study, is yes, with some important caveats. Coastal managers, like professionals in related fields, are embracing the new information technologies for their value in helping understand increasingly complex coastal problems and in formulating solutions. At the same time, they are awash in data and information of every imaginable type, value, and quality, delivered to their desktop computer at the click of a button, often whether they want it or not. They often do not have the time, technical capacity, or training to sort through, interpret, and apply much of the information that is available. But coastal managers want and need that capacity if they hope to be successful in addressing today's coastal problems.

Coastal managers are keenly aware that the problems their predecessors were addressing in the 1970s and 1980s were only the obvious ones—curbing the physical destruction of coastal habitats, treating industrial and municipal wastes, providing for greater public access to the shore, and so on. Today, there is greater scientific understanding and documentation of the complexities and interrelationships of coastal problems, and as layers are peeled away, there is also an appreciation for what we do not know. Finding solutions to problems that are acceptable to all is also difficult, requiring sophisticated approaches that bring together the best available scientific and technical information, employ the most effective information technologies, and make skillful use of communication and stakeholder processes. Adding to the challenge are increasing trends in key coastal development indicators—such as population density and change, coastal recreation demand, and investment in public infrastructure and private development—presenting ever-greater challenges for creative coastal management solutions. Reports of two recent blue-ribbon commissions, the Pew Oceans Commission (2003) and the U.S. Commission on Ocean Policy (2004), clearly document the continued ecological decline—indeed crisis—facing our coasts and oceans. Our management principles, institutions, practices, capacity, and political will to squarely face the problems are all insufficient to reverse the decline, the reports say.

Given this environmental dilemma, it is clear that effective documentation, sharing, and adaptation of best coastal management practices—whether innovative or simply tried and true—is an essential element in the long-term effort to create more sustainable futures for coastal communities and ecosystems. This study addresses that issue.

IMPETUS FOR AND SCOPE OF THE STUDY

In 2002, the National Oceanic and Atmospheric Administration (NOAA) Office of Ocean and Coastal Resource Management (OCRM) requested that The Heinz Center undertake a study of how innovative ideas and practices are shared within the coastal management community, and recommend improvements that would foster more effective and efficient information transfer. OCRM's interest in this issue stems from its responsibility under the 1972 Coastal Zone Management Act (P.L. 92-583) for administering the National Coastal Zone Management Program (Box 1.1). Today, that program includes thirty-four state and territory coastal management programs covering more than 99 percent of the U.S. coast-

---

Box 1.1   Technical Assistance Responsibilities of the Office of Ocean and Coastal Resource Management

Technical assistance to states developing and administering approved state coastal management programs has been an important function of the National Oceanic and Atmospheric Administration coastal management agencies since passage of the original Coastal Zone Management Act in 1972 (P.L. 92-583). Most recent amendments require that technical assistance be directed toward nine national interest areas identified in the Coastal Zone Enhancement Grants section of the law (16 U.S.C. § 1456b). Examples include coastal wetland loss, natural hazards impacts, cumulative impacts of development, aquaculture facility siting, and so on. Regarding these issues, the law states "The Secretary shall conduct a program of technical assistance and management-oriented research necessary to support the development and implementation of State coastal management program amendments under the Coastal Zone Enhancement Grants section of this title, and appropriate to the furtherance of international cooperative efforts and technical assistance in coastal zone management" (16 U.S.C. § 1456c.).

line, and a network of twenty-six state-administered National Estuarine Research Reserves (NERRs). OCRM administers grants to the states, monitors and evaluates their progress toward goals, and provides technical assistance.

Given its multiple responsibilities for more than thirty state programs that have both shared problems (e.g., managing shared estuaries) and common problems (e.g., managing development in hazardous areas), OCRM is often the first place state coastal managers turn to for advice. OCRM found that they often were unable to provide the kinds of technical assistance needed by states, either because of limited staff expertise or because they had no systematic way to document and make such information available (Douglas Brown, Deputy Director, OCRM, personal communication, June 28, 2003). Very simply, the agency has not been able to provide the technical and policy assistance their principal constituents need.

Further impetus was given for this study by findings from "customer service surveys" conducted by the NOAA Coastal Services Center (CSC), a national technical services and training center designed to provide many of the services states and other coastal managers are demanding from OCRM and NOAA generally. Although the CSC has filled an important niche as an information provider and broker, their constituent surveys suggest that the several hundred coastal managers they surveyed continue to be frustrated by their inability to obtain the information they need. In the 1999 CSC survey, 80 percent of those responding said they needed better access to information about how other agencies address similar issues (NOAA Coastal Services Center 1999a). Specifically, coastal managers found accessing and sharing information difficult, due to impediments such as communication gaps, lack of technical expertise, and past exaggerations about the promise of new technology. The CSC's 2002 survey findings were similar (NOAA Coastal Services Center 2002).

The scope of this study extended well beyond NOAA coastal programs. In this report, coastal management is defined broadly as the collection of strategies, policies, processes, practices, tools, and information used to make decisions that preserve, protect, conserve, and where appropriate develop and restore the coastal zone and its resources. It follows, then, that coastal managers—the audience—are the local, state, and federal agency decision makers and professional staff charged with implementation of the variety of laws and programs affecting the land, water, resources, and people who live, work, and visit coastal areas. Employees and members of nongovernmental organizations—environmental groups,

Box 1.2   What Do We Mean by Innovation and Diffusion?

Innovation and diffusion are two terms used throughout this report that bear definition, not because they require precise understanding, but because they are often subject to unnecessary debate. The committee borrowed the very simple but appropriate definition of innovation from Everett Rogers' classic text, *Diffusion of Innovations* (1995): "an innovation is an idea, practice, or object that is perceived as new by an individual or an organization." This definition was chosen for its breadth. Many of the practices, processes, and tools in coastal management are clearly offspring from other disciplines and times, yet they are new and innovative to the new user. For example, the "watershed approach" to management, clearly a new idea to many (i.e., an innovation) in the 1990s, was the same idea that John Wesley Powell was promoting in the 1890s as an environmentally sound organizing framework for county boundaries in newly forming western states (Stegner 1954).

The term *diffusion* is often interchangeable with information sharing. Rogers (1995) defines diffusion as "the process by which an innovation is communicated through certain channels over time among members of a social system." This report explores the importance and roles of different communication channels used in coastal management (e.g., face-to-face, conferences, the Internet), as well as the makeup and maturity of the various social systems or learning networks that encompass coastal managers.

trade organizations, and other interest groups—as well as private corporations and small businesses operating in or affecting the coastal zone also fit within this broad definition of coastal manager. The committee sought to learn from this diverse group and to develop recommendations for improving the sharing of innovations and the access to and diffusion of information for more effective coastal management (Box 1.2).

STUDY METHODS

In late 2002, The Heinz Center appointed a committee to explore how innovative ideas and practices for coastal problem solving are shared with others who have similar needs and how access to this information might be improved. Committee members were drawn from academia, government, industry, and environmental groups. At its initial meeting in

February 2003, the committee refined the overall study objectives and identified data that were needed to understand the issues and suggest solutions. Three principal objectives were identified:

- *Objective 1*: To define the problem more clearly by documenting how we now share innovative and other coastal problem-solving ideas and practices in government, academic, business, and not-for-profit sectors.
- *Objective 2*: To identify the strengths, limitations, and outlook for present information-sharing methods and efforts.
- *Objective 3*: To identify ways to improve information sharing and learning, drawing on the experiences of those within and outside the coastal management community.

Three data-collection efforts were initiated to address these objectives: (1) structured interviews with coastal managers, administered by committee members; (2) a workshop for coastal managers, designed to explore in more depth questions raised in the survey; and (3) identification of examples (in boxes throughout this report) illustrative of innovative coastal management practice and information-sharing techniques. Committee members also reviewed the literature and the World Wide Web for information about learning and communities of practice.

INTERVIEWS WITH COASTAL MANAGERS

Following its first meeting in February 2003, the committee developed a structured interview process designed to define the nature and extent of the information-access problem, to find out how coastal managers access information now, and to learn what improvements are needed for the future. The full set of questions are provided in Appendix A; the four main questions were:

1. What are the most important changes or improvements in U.S. coastal management that affect how you do business?
2. Where did you turn for information, ideas, strategies, or tools to help design and implement coastal management changes and improvements? What are the most important and credible channels of communication?

3. Generally, how do the really good ideas and practices in coastal management spread or diffuse?
4. How would you like to get information in the future, given constraints and resources available? What should be emphasized?

Thirty-five interviews were conducted with coastal managers from four federal agencies, ten state coastal management agencies, four state-administered National Estuarine Research Reserves, five local National Estuary Projects, three university-based Sea Grant extension programs, three ports, three environmental groups, and three scientific laboratories (Appendix B). Interviewees served a variety of coastal management functions and some more than one. Two were members of the U.S. Commission on Ocean Policy, six were primarily educators, six were researchers, three were involved in regulatory programs, six were planners, and four promoted coastal development and trade.

Neither the selection process for interviewees nor the responses to the interview questions provide the basis for an objective, scientific analysis of how coastal managers get and use information on innovative practices.* However, the interviews confirmed the premise of the study and educated committee members about the nature and extent of communication problems within the diverse network of professionals known as coastal managers. Furthermore, interviewees included coastal managers from business, environmental groups, government, and academia who may not be included in the NOAA surveys.

WORKSHOP ON COASTAL MANAGEMENT INFORMATION SHARING

The committee organized a two-day workshop in June 2003 to explore in more depth what was learned through the interviews. The workshop participants included many of the telephone interviewees and others they or committee members suggested. There are several reasons this group of seasoned coastal managers was assembled. First, the committee learned

---

* NOAA Coastal Services Center conducts statistically significant surveys of coastal resource managers every three years to determine technology and other training needs (see NOAA Coastal Services Center 2002 and http://www.csc.noaa.gov/survey/ for latest survey results).

from the interviews and knew from their own experience that a small workshop, face-to-face format was a powerful learning tool. Second, the committee wanted validation, elaboration, and feedback on their interview results. Key workshop panels focused on (a) how coastal information is being shared today, (b) how professionals in other fields are working to build more effective "communities of practice," and (c) how to use technology more effectively to get best practices and information to those who need it. The workshop also included breakout groups to give all participants an opportunity to weigh in on these questions. Finally, participants were asked what recommendations might be most useful to the coastal management community.

2

INNOVATION AND INFORMATION SHARING TODAY

FOR DECADES before the emergence of coastal management as a distinct practice, the coasts of the United States were managed—or mismanaged, many would say—through a number of single-purpose, often-conflicting policies and programs. The gradual transition to a more integrated, area-based approach to coastal management was spurred mightily in the 1960s by the work of the blue-ribbon Stratton Commission and its landmark report, *Our Nation and the Sea* (1969). At the same time, growing environmental activism was involving more and more people in what has become a movement toward a more participatory, transparent decision making process. Private sector awareness of the need to address environmental issues directly was also growing.

The Stratton Commission report and the transition toward more integrated strategies for addressing coastal issues were rooted in two competing views of coasts and oceans. One view held that the coastal zone's abundant natural resources and assimilative capacity for wastes from an increasingly productive society should be tapped to its fullest potential, while simultaneously protecting the environment. The competing view held that coastlines, estuaries, and oceans were already showing signs of severe stress due to unrestrained growth, dumping of wastes, and resource exploitation. These different perspectives persist today. Other Congressional, academic, and nongovernmental organization studies in the 1960s on recreation, estuaries, and coastal pollution further highlighted coastal environmental crises and resource development opportunities. Together, these efforts led to some of the most enduring national and state legislation for management of the coasts, oceans, and their natural resources. At the national level, these include the National Environmental Policy Act of 1969 (P.L. 91-190), the Coastal Zone Management Act of 1972 (P.L. 92-583), sweeping amendments to the Clean Water Act of 1972 (P.L. 92-500) and the Clean Air Act of 1970 (P.L. 91-604), the Marine Mammal Protection Act of 1972 (P.L. 92-522), the Marine Research, Protection, and Sanctuaries Act of 1972 (P.L. 92-532), the Fisheries Conservation and Management Act of 1976 (P.L. 94-265), and amendments to the Outer Continental Shelf Lands Act of 1978 (P.L. 92-629). These laws and programs have endured and evolved as needs have changed.

At the state level, the practice of coastal zone management was being defined in the late 1960s and early 1970s based on pioneering legislation in California (1965 and 1972), New Jersey (1970), Washington (1971), Rhode Island (1971), Oregon (1971), North Carolina (1974), and in other states and territories. Just as happened nationally, states also enacted or updated single-purpose legislation to address specific issues, such as beach access, shoreline development, wetland protection, and energy facility siting. Collectively, these new state and federal mandates and accompanying financial resources for planning, regulation, acquisition, and management resulted in significant innovation. Professionals charged with implementing or responding to these programs invented new management practices or tailored existing ones to their needs. Examples of these innovative coastal management practices and how they spread are provided throughout this report (see boxes).

Although states led the way with significant coastal management legislation, and other federal programs are important, it is the federal Coastal Zone Management Act of 1972 (CZMA), as amended, that provides the most comprehensive national framework for managing lands and waters of the nation's coastal zone. The CZMA sets forth broad policy guidance, declaring a national policy to "preserve, protect, develop, and where possible, to restore or enhance, the resources of the Nation's coastal zone for this and succeeding generations" (16 U.S.C. § 1451). Congress also provided federal financial resources and legal tools as incentives to states wishing to upgrade their capacity for coastal management. States first went through a planning and program development phase and then submitted a program proposal to the administering federal office for approval. Most state programs were approved in the late 1970s and early 1980s, and several states that dropped out early on have since developed programs and had them approved. Today, thirty-four states have approved programs covering more than 99 percent of the U.S. coastline.

Central to coastal management practice in the United States are the programs and activities at all levels of government that directly link to the federal CZMA and its policies. Under the CZMA, all three levels of government—federal, state, and local—have important roles to play and considerable flexibility in defining those roles. Although this flexibility has resulted in considerable diversity in state programs, the many issues states have in common or share have presented ample opportunities for adaptation of particularly innovative management practices from state to state and even coast to coast. How and to what extent the best of these practices get shared with other coastal managers and whether there is significant room for improvement in the sharing of innovative coastal management practices is a premise of this study.

COASTAL ENVIRONMENTAL ACTIVISM

While commissions and expert panels were influential in motivating the enactment of state and federal coastal and ocean legislation, there can be no doubt that grassroots activism played and continues to play a key role in coastal management. Thousands of teach-ins and marches associated with the first Earth Day (April 22, 1970) and the alarms sounded by Garrett Hardin (*The Tragedy of the Commons*) and Rachel Carson (*Silent Spring* and *The Sea Around Us*) helped to inspire and inform the formation of national, regional, and local organizations dedicated to protecting specific resources or challenging particular projects and policies that were seen as detrimental to the environment. These grassroots organizations have been empowered by legislation recognizing the role of citizens in shaping policy. Public disclosure ("sunshine") laws, citizen suit provisions, local participation opportunities, and other mechanisms such as direct citizen initiatives and lobbying for or against legislation through political action groups are increasingly important means of influencing policy. A recent example was the defeat in 2000 of a proposal to construct Dark Hollow Dam on Neshaminy Creek, a tidewater tributary of the Delaware River in Pennsylvania's coastal zone. The Natural Resources Conservation Service (NRCS) proposed to build a fifty-six-foot dam to help control costly flooding in the basin. The proposal was well down the road to construction when local citizens and environmental groups rose up against it.

Their opposition resulted in a more environmentally sound flood-control strategy that relied on voluntary buyouts, floodproofing of structures, and floodplain restoration.

A kind of "federalism" is at work among nongovernmental organizations, as well as governments. At a national or international level, the Natural Resources Defense Council, Environmental Defense, Greenpeace, Sierra Club, and other broad-purpose environmental groups include ocean and coastal issues among their areas of concern. Wildlife management groups, such as the Isaac Walton League and Ducks Unlimited, also work to conserve coastal species and habitats. The Coast Alliance, Ocean Conservancy (formerly the Center for Marine Conservation), and Oceana are some of the largest and most active national groups focusing primarily on oceans and coasts. In addition, many regional, estuary-specific, or state organizations undertake the protection, restoration, or management of particular coastal areas or watersheds, such as Heal the Bay (Santa Monica), the Northwest Straits Marine Conservation Initiative (Washington State), Save the Bay (Narragansett), the Chesapeake Bay Foundation, and the Gulf of Mexico Restoration Network. Nested within many of these larger organizations are groups that focus on a particular shoreline area or tributary of an estuary (e.g., Friends of the Elizabeth River, James River Association, and Riverkeeper).

THE PRIVATE SECTOR AND COASTAL MANAGEMENT

The private sector and the marine industry in particular were major players in the early debates and development of marine and coastal policy. Although environmental protection was an important theme in the 1969 Stratton Commission report, the need for the country to develop marine resources to their full potential was even more central to the report. Important themes included the strategic importance of marine transportation, the need for expansion and modernization of seaports, and the potential of the outer continental shelf as a source of oil, natural gas, and other minerals and of the deep seabed for strategic minerals. The need to develop and Americanize marine fisheries was another important issue. All were central to early debates on ocean and coastal policy.

Organizations such as the American Association of Port Authorities (AAPA), the American Petroleum Institute (API), and the National

Ocean Industries Association (NOIA) represented development interests and pressed for "balance" between conservation and development in emerging coastal environmental policies and programs. Coastal environmental activists and industry interests were generally at opposite poles, with government agencies on one side or the other (depending on missions) or straddling the fence, trying to build consensus. As might be expected, the courts were often called upon to mediate or arbitrate contentious issues.

As new policies were being debated, forged, implemented (and contested), more enlightened business interests began reacting to a new awareness of the need for proactive approaches to environmental issues. A wide variety of business activities aimed at environmental stewardship have sprung out of the awareness that to stay in business, business must share society's goals. Thus, today there are many business initiatives that are designed to demonstrate corporate responsibility and put corporate resources to work preserving and enhancing the environment, including coasts and oceans. For example, business has embraced and actively participated in both the establishment and support of National Estuary Programs (NEPs) throughout the country; particularly in Texas, NEPs have strong business involvement and leadership. In Washington State, the energy industry and ports are heavily involved in the Northwest Straits Marine Conservation Initiative, resulting in new industry–government–environmental partnerships.

The United States Business Council for Sustainable Development (USBCSD) grew out of the Gulf of Mexico Business Council for Sustainable Development, which was established in 1993. The original group was started in part to complement the establishment by the U.S. Environmental Protection Agency (USEPA) of the Gulf of Mexico Program. Launched in 2002, the USBCSD is a partner organization of the World Business Council for Sustainable Development (WBCSD), a global network of 160 international companies with members drawn from thirty countries and twenty major industrial sectors. The WBCSD plays a leading role in shaping the business response to the challenges of sustainable development. The USBCSD plays a complementary role by communicating those policies to the U.S. business community and its stakeholders, and by implementing projects that apply sustainable development principles to real-world problems.

The Wildlife Habitat Council (WHC), established in 1988, helps large landowners, particularly corporations, manage their unused lands in an ecologically sensitive manner for the benefit of wildlife. WHC also works to broaden understanding of wildlife values. Over 120 companies are members of the WHC, including two dozen conservation organizations, plus many supporters and contributors. Over two million acres in forty-eight states, Puerto Rico, and fifteen other countries are managed for wildlife enhancement through WHC-assisted projects. Many of these are in coastal areas.

Another example of business activity in habitat preservation and enhancement is The Nature Conservancy, which has been at work since 1951. The International Leadership Council (ILC) of The Nature Conservancy is one of the world's leading corporate forums focusing on the challenges confronting biodiversity preservation, habitat conservation, and natural resource management. These issues lie at the heart of a growing number of corporate responsibility programs. The ILC brings together companies from many industries—finance, manufacturing, forestry, consumer products, information technology, etc.—to seek solutions to conservation challenges through cooperative partnerships between the business community and The Nature Conservancy. ILC members are global companies with a strong commitment to finding innovative ways to promote sustainable development and biodiversity preservation. Its thirty-six members each contribute more than $25,000 annually. A new Nature Conservancy "marine initiative" is extending this cooperation to offshore environments.

National Fish and Wildlife Foundation (NFWF) corporate partnerships also allow business to use financial resources for environmental preservation and enhancement, including in the coastal zone. The NFWF supports innovation and nontraditional approaches to forge sustainable partnerships between private landowners, corporations, natural resource agencies, and conservation organizations. Among its fifteen corporate partners are coastal management–related programs with significant contributions from Shell, FMC, Conoco Phillips, and the Southern Company. Since its inception in 1998, the Shell Marine Habitat Program has catalyzed priority conservation around the Gulf of Mexico and elsewhere. Shell's funds have provided much-needed and significant private sector funding that has leveraged millions of additional private and public dollars. By providing "venture capital" for projects in the Gulf, the Shell Marine Habitat Program has revitalized partnerships in the region and established clear leadership for the corporate sector in the conservation arena. To date, ninety-two projects have been funded with a total conservation impact of nearly $16 million.

These are but a few examples of private sector initiatives that are proactive in the preservation, restoration, and protection of coastal and ocean resources. The private sector is also in the vanguard in other environmental initiatives with important coastal management implications, among them "green" movements for hotels (see Box 3.3, pages 58–59) and marinas, and the "smart growth" movement focused on creating more livable, environmentally healthy communities.

Nevertheless, tensions between private interests, government, and environmental activists persist and are reflected in both positive and negative reactions to recommendations of the Pew Oceans Commission report (2003) and the preliminary report of the U.S. Commission on Ocean Policy (2004). These reports, both of which call for a more coherent national ocean policy, governmental restructuring, new and more drastic actions to protect marine fisheries and coastal habitat, and tackle difficult pollution problems, no doubt signal a new round of debates on complex, high-stakes issues at the coast.

WHAT DRIVES CHANGE IN COASTAL MANAGEMENT PRACTICE?

Exploring what drives change in coastal management practices was one of the first priorities for the Heinz committee, to better understand how innovations and information diffused within the community. One question of interest was the importance attributed to these changes by coastal managers—our interviewees and workshop participants. Our first interview question (Appendix A) dealt with this directly by asking, "Looking back over the last decade or so, what are the most important changes or improvements in coastal programs, planning and development processes, regulatory or acquisition programs, etc., that have affected how you do business?" We followed up with a question about the motivations or drivers underlying these changes. The results can be classified into several categories of change drivers: social and economic change; improved scientific understanding of problems and application tools; advances in information technologies and tools; policy improvements, initiatives, and financial incentives; improved planning and management processes; and outreach and technical assistance (Table 2.1). In each of the areas where coastal managers reported significant changes that affected how they

Table 2.1   Drivers of Change and Changes Affecting U.S. Coastal
Management Practice, as Reported in Committee's
Interviews and Workshop

Drivers of Change	Changes Affecting Coastal Management Practice
Social and economic changes	Growth of environmental activism, public participation, and the citizen initiative movement Population growth at the coast, permanent and visitor Expanded investment and redevelopment of the coast Port consolidation, shift toward tourism economies on rural coasts
Scientific understanding and tools	Increased interdisciplinary research applied to management Expanded definition of the "coastal zone" inland to watersheds and offshore to marine realm (e.g., nonpoint source issue) Tools for monitoring change (e.g., NOAAs CCAP, LIDAR, ocean-observing systems, etc.)
Technological advances	Explosion of Web-available data and information More rapid communication—e-mail and data transfer Growing GIS capacity—mapping, analysis, communication Metadata tools for documenting and assessing data quality

Policy initiatives and incentives	Broad paradigm shifts
	  Integrated coastal management
	  Sustainable development
	  Ecosystem and watershed approaches
	  Community-based planning
	  Species and habitat restoration
	National policy initiatives
	  National Estuary Program
	  Nonpoint source pollution (CZMA 6217, Clean Water Act 319)
	  Coastal Zone Program enhancements (CZMA 309)
	  Pew Oceans Commission, Oceans Act, and the U.S. Commission on Ocean Policy
	  Marine protected areas
	  Urban coasts focus—brownfields, revitalization, restoration
	  Legal decisions (e.g., takings, access, public trust)

Planning and management processes	Increased intergovernmental and intersectoral collaboration Greater transparency in decision making processes Improved engagement of stakeholders Incorporation of consensus building and alternative dispute resolution into processes
Outreach and technical assistance	NOAA Coastal Services Center technical assistance and training programs Sea Grant Coastal Training Programs—National Estuarine Research Reserves

CCAP, Coastal Change Analysis Project; CZMA, Coastal Zone Management Act; GIS, geographic information system; LIDAR, light detection and ranging; NOAA, National Oceanic and Atmospheric Administration.

do their jobs, innovations in both practices and information sharing played important roles in the process.

SOCIAL AND ECONOMIC CHANGES

Social and economic changes, particularly the steady growth in permanent and visitor populations at the coast and even more rapid growth of investment in development and redevelopment, was cited as a major driver of coastal change—the "elephant in the living room," according to one workshop participant. Changing economic forces—from local to global—were another driver cited, examples being port consolidation and the growth of coastal tourism in rural areas. More informed and sophisticated public participation in planning and management, and its institutionalization in most coastal programs, was cited as another important change.

SCIENTIFIC UNDERSTANDING AND TOOLS

Increasing sophistication of coastal and marine science was cited by many as an important change in the past twenty years, providing a more certain (but still disturbingly uncertain) understanding of the management challenges we as a society face. One example noted was the increasing incidence (or awareness) of biological "dead zones" along our coasts—the best known and most publicized one being in the Gulf of Mexico offshore from the Mississippi River delta. The limits of science were also noted, with collapses of "scientifically managed" fisheries an example.

Participants in the committee's interviews and workshop also made special mention of the importance of the specialized research and technology programs designed to feed new and more sophisticated information into coastal decision-making processes. Examples noted include the research programs of the National Estuarine Research Reserves (NERRs), and the new technology development institute the National Oceanic and Atmospheric Administration (NOAA) has established at the University of New Hampshire—the Cooperative Institute for Coastal and Estuarine Environmental Technology, or CICEET (Box 2.1).

**Box 2.1   CICEET: Developing Innovative Technology for
Coastal Monitoring**

The Cooperative Institute for Coastal and Estuarine Environmental
Technology (CICEET) was established in 1997 as a national center for
research, development, and application of innovative environmental
technologies to address pressing environmental issues in estuaries
and coastal waters. Located at the University of New Hampshire
(UNH), CICEET is a unique partnership between UNH and the National
Oceanic and Atmospheric Administration (NOAA). CICEET is jointly
managed by UNH and NOAA co-directors and uses the capabilities of
federal, state, and local government, the private sector, research
institutions throughout the United States, and the twenty-six
National Estuarine Research Reserves.

What would eventually become CICEET began as a concept for
a new approach to problem solving in the coastal environment
advanced by biologists, estuarine ecologists, oceanographers, and
environmental engineers at UNH. This group recognized that development pressures and changing demographics in U.S. coastal areas
will continue to present challenges to coastal managers, who will
need new tools to deal with these issues. Committed to a research
and development program that would directly improve the way
coastal resources are managed, the group established several principles to guide the program:

- Individuals and groups that are directly engaged in making management decisions that affect coastal resources are best equipped
  to identify priority issues.

- Research and development needs to be directed at finding solutions to priority problems.

- Today's problems are complex and require a multidisciplinary, collaborative approach.

- New environmental technologies and techniques can provide solutions to some of these problems, much as they helped to reduce
  pollution from point sources.

- Tremendous advances in information technology and communications can transfer environmental technologies to users quickly and
  efficiently.

Applying these guiding principles, UNH focused on research and
development of environmental technologies that would improve
control of pollution and lessen degradation of the coastal environment. NOAA's reaction to the UNH proposal was positive; in fact,
NOAA urged UNH to consider applying this approach throughout
the country, using the twenty-six National Estuarine Research Reserves

Box 2.1 continued

as the field platforms for technology development and demonstration. The result was an agreement between NOAA and UNH establishing CICEET as a national program that would be jointly managed based on the shared goals and operating principles.

CICEET's approach has been to identify the most critical problems, focus competitively awarded research and development on innovative solutions, and identify the most effective means of informing those who are responsible for implementing change. To identify and prioritize problems, CICEET works closely with organizations such as the Coastal States Organization, the National Estuarine Research Reserve System (NERRS), NOAA's National Ocean Service, and the USEPA's National Estuary Program. Using their input, CICEET focuses its applied research programs on developing new technologies and approaches to problem solving. Each year, CICEET solicits proposals for research on topics identified as high priorities by coastal managers. Broad distribution of solicitations and a rigorous selection process mean that the best science and technology can be brought to bear on environmental problems. Since the program began in 1997, more that 100 projects have been supported at NERRS field sites throughout the country. To make technology and information available to coastal managers, CICEET organizes issue-oriented workshops, special sessions at national meetings, Internet-based project bulletins and spotlights, training for new grant applicants, and a new program to develop and implement project-specific technology transfer strategies. Although CICEET is still new compared with other pollution abatement and prevention efforts, it has already developed technologies and methods and launched projects that hold great promise for effecting positive environmental change in the coastal environment.

TECHNOLOGICAL ADVANCES

New technology, such as the emergence of the World Wide Web as an information-sharing innovation, was noted by virtually all of the interviewees as the preeminent change affecting how they worked. Google, the powerful Internet search engine, is attaining status as a *verb*, some noted. Others suggested that a tailored "coastal management google," enhanced by a human interpreter or user ratings like those employed by Amazon.com, would solve many of their information-gathering needs. Many managers identified the availability of remote sensing and geographic information system (GIS) applications, along with access to geospatial data, as another significant technological advance.

Policy Initiatives and Incentives

New policy initiatives, particularly when accompanied by funding and other incentives for participation, often are important catalysts for change in coastal management. The Coastal Zone Enhancement Grants program (CZMA 309) for example, provided a new pot of funds for states to analyze the strengths and weaknesses of their coastal programs and to develop strategies for improvement, many of which were quite innovative. Another example cited by several of our interviewees was the National Estuary Program, established in 1987 as an amendment to the Clean Water Act. This program embraced and did much to advance the watershed approach to coastal and estuarine management (Box 2.2). Some policy changes, however, have been judged less successful—the non–point pollution control amendments to the 1990 Coastal Zone Act Reauthorization Amendments, for example—where initially very little funding was provided. Legal

---

Box 2.2   The USEPA's National Estuary Program

States and local governments have clamored to take part in the National Estuary Program, authorized as part of the Water Quality Act of 1987 (WQA). This legislation was enacted following extensive testimony, debate, and modification over a period of two years. WQA added Section 320 to the Clean Water Act, which provides incentives, including grants, for collaboration among agencies at all levels of government. Five rounds of nominations elicited more than forty nominations from state governors for estuaries to be included in the program, of which twenty-eight were designated "national estuaries" by the USEPA administrator. Each of these partnerships characterized conditions and threats to the ecological integrity of the designated estuary and completed "Comprehensive Conservation and Management Plans," which were approved by the USEPA, to guide restoration efforts. The Association of National Estuary Programs worked with Congress and the USEPA to secure a modification to the Clean Water Act that allows the USEPA to award grants for the implementation of approved management plans, eliminating a limitation in the statutory language that had originally provided only for grants to "develop" such plans. The USEPA continues to receive expressions of interest in nomination of additional estuaries. The National Estuary Program model proved so popular and successful that it helped inspire the USEPA's broader movement to watershed management, begun in the early 1990s.

decisions, such as the Lucas "takings" case in South Carolina and numerous other court decisions, were noted as important drivers as well. Finally, the impact of policy changes that affect coastal pressures in other countries and at the global level was noted.

PLANNING AND MANAGEMENT PROCESSES

Improved planning and management techniques were another important area of change in coastal management. Examples include the incorporation of consensus-building and dispute-resolution techniques into coastal planning, the institutionalization of stakeholder representation in decision making, and a greater emphasis on understanding and overcoming barriers to the effective use of science in management (Box 2.3).

Box 2.3   Industry Stakeholders in the Gulf of Mexico Program

The Gulf of Mexico Program is an innovative example of representatives of private organizations becoming full partners in a government environmental initiative.

The Gulf of Mexico Program was established by the U.S. Environmental Protection Agency in 1989 in response to a strong regional citizens' movement to create a program similar to the Great Lakes and the Chesapeake Bay Programs. The program's initial focus was on coordination of the various scientific and research programs in the Gulf of Mexico ecosystem. Its initial emphasis was on organization of federal and state agency efforts. There were also several Congressional initiatives to establish the program in statute, some moving the lead role into the National Oceanic and Atmospheric Administration. One component of the original program was a Citizens' Advisory Committee that was to represent five different segments of stakeholders. However, the governors of the various states, creating an inherent political bias, appointed the committee. For example, Texas had at one time a plaintiff lawyer representing business and an environmental regulatory agency employee representing tourism. This committee failed to demonstrate it could effectively represent a variety of stakeholders' interests in implementing necessary program initiatives.

With the support of the Management Committee (the implementation body of the program that met regularly), program staff drafted a set of criteria for seating interest groups on the Committee.

Box 2.3   continued

The Management Committee evaluated the draft criteria and ultimately endorsed seating outside (nongovernmental) groups provided the groups met the approved criteria. The first group to meet
the criteria for inclusion as a partner was the Gulf of Mexico Business
Council. This group also successfully petitioned for its inclusion on
the Policy Review Board after being seated on the Management
Committee. The Business Council was organized through the pre-
existing Gulf of Mexico Coalition (a business group organized to
interface with the Gulf of Mexico Program).

The key to the program's stakeholder involvement is full participation as partners. Full partnership ensures that the quality of decisions
and efforts are maximized. The program has the opportunity to focus
on issues that the variety of stakeholders and governmental agencies
can agree to pursue. Once a decision is reached, political opposition to
implementation is minimized. Of course, there are costs that go with
such involvement. Decisions take longer to make and sometimes the
preparation for a decision carries a higher financial cost.

Since the addition of the Business Council, the program has
added representation from the Gulf Restoration Network (environmental advocacy), the American Farm Bureau–Gulf Chapter (agriculture), and the Conference of Southern County Associations (local
government). The program has also recently united in a decision to
request a Presidential Executive Order establishing the program in
a more formal manner with a higher governmental profile. The
request came from the governing body, where the stakeholder participants play a pivotal role. To many, this effort may determine
whether the whole experiment of stakeholder partnership will bear
fruit; but to others, the collaboration that has followed this organization itself is a model. The very fact that business representatives
are full participants in a governmental program is something that
community cherishes. But, some other participants are more reserved
in their judgment (Gulf Restoration Network). Efforts to share this
innovation have been limited to the business community in the Gulf
States. While governmental representatives have marveled at the
level of stakeholder cooperation in the program, to this point only
the business community and the program itself seem interested in
diffusing the way such a cooperative attitude was reached.

When queried, the program office shared some of its lessons
learned. First, the program suggests that establishment of criteria for
involvement in any such partnership arrangement should only be the
start of an effective partnership. The criteria should be allowed to continually evolve to better match the stakeholders' expectations. And, for
the collaboration to bear fruit, there needs to be an effective method
for developing shared accountability. Thus, expectations of participants
are an essential part of any criteria for participation. But again, these
must be subject to input and consensus from the potential participants.

OUTREACH AND TECHNICAL ASSISTANCE

Outreach and technical assistance efforts for coastal managers have also matured and improved in recent years. Many cited the 1994 establishment of the NOAA Coastal Services Center as one of the most important changes in technical assistance and outreach, particularly the center's work to advance the use of GIS and build local capacity for its practical application to coastal problems. Others cited Sea Grant programs as vital supporters of state and local coastal programs, both through their university-based research programs and their local outreach efforts. The Coastal Training Program for local decision makers, which is being advanced by National Estuarine Research Reserves, was another highly touted program. These three often-mentioned programs are described in some detail in Box 2.4.

Box 2.4   Information Purveyors in Coastal Management

**SEA GRANT'S INFORMATION TRANSFER MODEL:
LINKING RESEARCH, EDUCATION, AND OUTREACH**

The National Oceanic and Atmospheric Administration (NOAA) Sea Grant Program, established in 1966, engages the nation's top universities and colleges to conduct research, educate students, and reach out to marine users and the public. One of Sea Grant's strengths is its extensive but well-distributed national network—there are thirty Sea Grant college programs across the country, with more than 3,000 participating scientists, engineers, outreach experts, educators, and students at some 300 institutions. Another strength is its approach to research and learning. Sea Grant scientists and outreach experts transfer scientific research results to industry and the coastal management community, provide training opportunities for K–12 educators and mentoring opportunities for undergraduate and graduate students, and help keep the public informed about marine and coastal issues. The program operates on a two-way communication model, with representatives of industry and other users of Sea Grant information providing feedback on research and outreach needs and priorities.

Coastal managers are an important partner and audience for Sea Grant research and outreach programs. To support coastal management, Sea Grant programs initiate and join partnerships; organize workshops, training, and demonstration projects; and publish extension bulletins, professional meeting and conference proceed-

Box 2.4 continued

ings, how-to manuals, Web sites, and newsletters relevant to coastal decision makers. A few examples of Sea Grant programs with important coastal management benefits include:

- *Project NEMO*: Nonpoint Education for Municipal Officials, originally a Connecticut program, that provides tools for improved water quality planning and development (see Box 4.5, pages 89–90).

- *Aquaculture for Regulators*: A program in Massachusetts to provide regulators with a better understanding of the shellfish industry.

- *Aquatic Nuisance Species Programs:* Programs in California, Oregon, Washington, and other states designed to engage ports, businesses, and local volunteers in prevention, detection, and control of invasions.

- *Volunteer Monitoring:* Water quality programs in many states (e.g., Estuary-net, http://www.ncnerr.org/education/estnet/) with schools and local watershed groups to assist in collection of data and evaluation of environmental indicators.

- *Coastal Planners Group*: An educational collaboration between Washington Sea Grant and the state Coastal Zone Management Program designed to bring the latest, relevant science on coastal issues to planners in Puget Sound (Goodwin and Canning, 2001).

- *Coastal Hazards Mitigation Programs:* Programs in many states educate the public, planners, realtors, builders, and others about mitigation strategies, construction and remodeling techniques (e.g., the 113 Calhoun Street Foundation, http://www.113calhoun.org/), sustainable land use, and more.

- *Waterfront Revitalization Programs:* In Virginia, Washington, and Oregon to assist smaller communities in planning for waterfront improvements.

The Sea Grant Program—which combines university-based research applied to issues of both local and national significance, outreach programs that put research findings in the hands of users and managers, and the communication of information needs back to program administrators—serves as an important model for the coastal management community.

THE NOAA COASTAL SERVICES CENTER: LINKING PEOPLE, INFORMATION, AND TECHNOLOGY

The NOAA Coastal Services Center opened for business in 1994 in a vacated naval facility in Charleston, South Carolina. Its beginning was modest, but the ambitious vision for what the center was to become was firmly in place.

Box 2.4 continued

NOAA wanted a technological catalyst for improving coastal management practice, an organization that could bring new and underutilized science and technology to the coastal resource management community. It also wanted an organization that focused on real, on-the-ground situations and looked to the customer to set its agenda. To do this, the center solicited project proposals and partnerships from its customers. By adopting an operating principle of "national in scope, local in approach," the center has been successful in resolving issues and applying the tools and lessons learned from each project to other coastal states and communities.

Customer input continues to guide the philosophy, areas of primary interest, and operating principles of the Center. All projects are customer focused and include appropriate partners. Evaluations are conducted to ensure that the product or service meets client expectations, to determine program effectiveness, and to guide future efforts.

Each product or service is created in response to a specific local issue, but also has a national component, so that lessons learned and technology harnessed or created for each effort are transferable to a larger audience. This "national in scope, local in approach" operating principle requires that the center not only work with individual clients, but that it has a good understanding of the national community as well.

Results from customer surveys, examples of which are discussed in this report, help center staff prioritize issues and create products and services compatible with the computer hardware and software programs favored by the majority of the coastal resource managers. The customer survey is repeated every three years. Survey results reveal where the customer base expects to be in the future in terms of management issues and technological capabilities. This information is shared with the rest of NOAA and the state coastal resource managers.

The NOAA Coastal Services Center's Web site gives snapshots of the kinds of assistance provided to coastal resource managers (http://www.csc.noaa.gov/), and a number of these are described in this report. What is clear from this study is that the center has become the "go-to" source for state coastal managers interested in applications of GIS, remote sensing, and other technologies, as well as other general training needs.

THE NATIONAL ESTUARINE RESEARCH RESERVES COASTAL TRAINING PROGRAM

With a goal of promoting more informed coastal decisions through science-based training, the National Estuarine Research Reserve System (NERRS) developed the Coastal Training Program nationwide (http://www.nerrs.noaa.gov/Training/welcome.html). Local and regional forums are created to bring the best available science-based infor-

Box 2.4 continued

mation, tools, and techniques to bear on decisions that affect coastal resources in watersheds, estuaries, and near-shore waters. The program capitalizes on several unique characteristics of the reserve system—local presence, knowledge, and partners—to provide workshops, seminars, distance learning, technology applications, and demonstrations aimed at coastal management professionals and decision makers.

Decisions made by coastal communities can have profound, long-term consequences for estuarine and coastal environments. Elected officials, land use planners, regulatory personnel, coastal managers, and agricultural and fisheries interests are key decision makers who often do not have adequate access to relevant science-based information, training, or to make informed decisions affecting the coast.

Opportunities for information exchange and skill training expand coastal management networks, increase collaboration across sectors, and improve local understanding of the environmental, social, and economic consequences of human activity within the coastal landscape. Reserve staff identify critical issues in the region and the key coastal decision makers that could benefit most from relevant science and training. Participants in the NERRS Coastal Training Program might include state and local elected officials, agency staff, volunteer boards, key members of nongovernmental organizations, business organizations, and state and regional professional associations whose daily decisions influence coastal resources.

The national network of twenty-six Reserves implement the Coastal Training Program partnership with a variety of national and local organizations. At the national level, NOAA's Estuarine Reserves Division (ERD) provides strategic and budget planning and support in partnership with NOAA's Coastal Zone Management Program, Sea Grant, and the Coastal Services Center. At local and regional levels, individual reserves develop local partnerships to ensure effective implementation. Key partners may include state coastal programs, state Sea Grant programs, local universities, research institutes, professional organizations, local government agencies, nonprofit organizations, and a variety of others with expertise, skills, training sites, and logistical support. Initially developed at one estuarine research reserve, the program expanded system-wide (twenty-six sites) in 2002 to increase the capacity of reserves to deliver technical training services to underserved constituent groups at the local level. Lessons learned include the importance of local knowledge and contacts to disseminate coastal management innovation and the importance of local presence and resources to provide follow-up support beyond formal training programs. The Coastal Training Program provides reserves with opportunities for economies of scale, increased efficiency, and targeted program development.

WHERE DO COASTAL MANAGERS
GO FOR IDEAS AND INFORMATION?

When coastal managers have a problem and they want to explore how others have dealt with similar situations, where do they turn for information? How do they learn about innovative or reliable practices for addressing the issue? And to whom do they turn first? These questions were central to both our interviews and the follow-up workshop. The answers provide insight into the nature and social structure of the coastal management community and the kinds of information-sharing improvements that might work for them. We also found that our results corresponded closely with those of the NOAA Coastal Services Center (2002) customer survey (http://www.csc.noaa.gov/survey/).

Coastal managers cast a broad net when seeking new information for problem solving (Table 2.2). Although they use a variety of sources, the most important and credible source varies with the issue. Despite the explosive growth of the Internet, managers still rely most heavily on their

Table 2.2 Methods and Sources for Gathering Information

How Coastal Managers Get Information	Most Credible Sources
Tap into personal experiences and networks—own expertise or colleagues in related organizations	People known and trusted and the people they know and trust
Multiplicity of sources—depends on problem or opportunity presented	Information purveyors: National Oceanic and Atmospheric Administration Coastal Services Center, local Sea Grant, the U.S. Environmental Protection Agency, Office of Ocean and Coastal Resources Management, Coastal States Organization, American Association of Port Authorities, university experts, professional societies, journals
Internet— Google and other search engines increasingly used (need to filter)	
Directed, systematic approach for clearly identified issues in work plans	
Serendipity often plays role—hear, read, or see something and recognize relevance	
Conferences (for networking) and workshops (targeted information) both important	
Stakeholder and advisory groups—local knowledge important for local solutions	Experts—people from out of town
Consultants used to gather and filter information	

personal network of contacts—people they trust. This finding was congruent with the findings of the NOAA Coastal Services Center's customer survey. When asked how they get or exchange new information, 98 percent of their respondents rated "talking with colleagues and friends" highly or moderately important (NOAA Coastal Services Center 2002).

Nevertheless, every person interviewed in this study acknowledged the rapidly growing importance of the Web and the Internet as a resource for learning what others have done to solve particular problems, and as a source of data, especially digital, spatial, and remotely sensed data. This trend is reinforced by the technology-savvy professionals entering the field, who make increasing use of tools like geographic information systems and sophisticated image-processing software to analyze data for resource management and decision making.

Many interviewees said that conferences (e.g., the biennial Coastal Zone conferences and the Coastal Society conferences) were most valuable for networking—building those personal relationships noted earlier. However, when it came to learning about new approaches to planning or management, targeted workshops were more important than conferences. These results also closely parallel the Coastal Services Center survey findings, which revealed that 90 percent of respondents rated conferences as either "highly or moderately important" in getting or exchanging information and that 85 percent rated workshops in the same high or moderate category (NOAA Coastal Services Center 2002).

State-level managers interviewed by the committee often cited particular agencies as important sources—usually agencies that sponsored their programs. For example, interviewees who ran National Estuary Programs cited the USEPA, while state coastal program managers more often mentioned the NOAA Coastal Services Center or the Office of Ocean and Coastal Resources Management. This is not surprising, considering the sponsor–client relationship inherent in these programs. Clients are wedded to their program sponsors for policy guidance, funding, technical assistance, performance evaluation, and other program implementation activities. Nevertheless, this "stovepipe" model may be inhibiting the flow of innovative ideas and information among programs with similar goals and may even set up a competitive situation, where organizations attach themselves to or become identified with particular innovations, much the same as private sector companies protect their proprietary interests in ideas. This will be explored later.

Sea Grant programs were another oft-noted information source, as were university experts and lobbying organizations, such as the Coastal States Organization and the American Association of Port Authorities. Again, interviewees cited those most strongly associated with their own program or function, but not peripheral organizations. One interpretation of this is that program sponsors and lobbyists for programs have done a good job of providing valuable services to their clients.

Professional journals and professional societies, such as *Ocean and Coastal Management, Coastal Management,* and the Coastal Society were noted by our interviewees, but generally deemed of less importance as information sources. This assessment differs somewhat from the Coastal Services Center's findings, where 69 percent rated scientific journals as highly or moderately important for information exchange (NOAA Coastal Services Center 2002). Finally, experts in the field were recognized as important resources. One state coastal manager noted that when faced with a new issue, he simply hires the most knowledgeable person available to do an exhaustive literature review, using that as a starting point for developing a unique approach.

HOW DO INNOVATIONS SPREAD?

Another area of inquiry for the committee was how, *in general,* the good ideas and practices in coastal management diffuse or spread. The views of our interviewees and workshop participants simply reinforced the committee's earlier, more personal observations (Table 2.3). Personal, face-to-

Table 2.3   Diffusion Mechanisms for Innovations

How good ideas get diffused to coastal managers
  Policy changes often drive innovation and diffusion.
  Champions at different levels are important.
  Availability of funds encourages experimentation and tailoring (e.g., CZMA 309).
  Personal contacts and face-to-face interaction spread ideas.
  Learning occurs through conferences and workshops.
  Young professionals bring in holistic approaches, knowledge, and technological sophistication.

face contact, networking at conferences, and learning at workshops were all confirmed as principal avenues for diffusion. An example of natural policy as a driver of innovative change is the 1987 amendments to the Clean Water Act, which established the National Estuary Program, which in turn helped spread the "watershed approach" to coastal resource management, engaging local residents and governments in on-the-ground problem solving. This policy driver was particularly important because funds were attached that provided the opportunity to take up a new approach.

Another important factor in the diffusion of innovations was champions, or what John Kingdon (1995) calls "policy entrepreneurs"— individuals or organizations that attach themselves to a new tool or approach and make it their mission to make others aware of its uses and merits. The special area management planning (SAMP) model is particularly relevant here. As described in Box 2.5, the efforts of participants and academic observers in Grays Harbor, Washington, in the mid-1970s led to amendments to national legislation and adaptation of their planning process by other agencies and programs—even common use of SAMP internationally.

Box 2.5   Special Area Management Planning: The Journey from Grays Harbor

Special area management planning, or SAMP, is a regional planning process that incorporates research, planning, and implementation techniques borrowed from dispute-resolution theory and experience. The process has been used in a variety of coastal environments to address habitat degradation and restoration, runoff pollution, natural hazards, port development, and other issues.

The SAMP process was first described in the late 1970s by those involved in planning for development and protection in Grays Harbor, Washington. Coastal planners there faced a variety of difficult issues—an internationally important migratory bird resource that was threatened by proposed port development, local economic decline and the need for diversification to create jobs, and a long history of regulatory conflict between the port, local governments, and state and federal agencies. Ultimately, under great political pressure to resolve long-standing conflicts, a planning task force was formed in 1975.

Working with a team of consultants, coastal planners brought together stakeholders to develop a consensus-based plan for water, wetland, and shoreline policy and zoning for distinct "management

Box 2.5 continued

units." After several years, millions of dollars, and several lawsuits, the plan was completed, adopted, and enacted. Implementation was through local CZM programs and state and federal regulatory processes. Since its approval—with regular updating—the plan has been the blueprint for development and resource protection in Grays Harbor.

The SAMP model that emerged from the Grays Harbor process described by Evans and colleagues (1980), has a number of key requirements and features:

- SAMP is a collaborative, multilevel, multisector process.
- Agreement is by consensus—this is necessary because all stakeholders have the power to make it work or fail.
- SAMP integrates authorities at all governmental levels.
- A neutral mediator runs the process, because mistrust is often a major problem.
- Affected groups and individuals are involved in meaningful ways.
- Implementation is considered throughout the process, with mechanisms specified and owned by stakeholders.

NOAA funded much of the Grays Harbor process and promoted the SAMP model in their 1980 Report to the President on Coastal Zone Management (NOAA Office of Coastal Zone Management 1980). SAMP was then adopted by Congress as national policy in the 1980 amendments to the Coastal Zone Management Act (CZMA 303). Policy support and funding from NOAA led many states to experiment with the SAMP process to address a variety of complex, regional coastal problems. This diffusion gained further impetus through the 1990 amendments (CZMA 309), when Congress required states to examine the potential of SAMP to improve their management programs and, if appropriate, use the tool. Today, some twenty states have developed one hundred SAMPs as part of their coastal management planning. Among the best described are those in Rhode Island, particularly the Salt Ponds SAMP (Olsen and Lee 1991; Imperial 1999). The Delaware Coastal Program has also made extensive use of the SAMP model, following a more elaborate process developed and promoted by NOAA (Delaware Coastal Program 2001). The SAMP process has also spread to other U.S. government agencies (e.g., U.S. Army Corps of Engineers) and internationally (e.g., Ecuador and other countries, assisted by the University of Rhode Island Coastal Resources Center's international program). The University of Rhode Island has also developed an international seminar on SAMP that it has delivered to more than 200 participants over ten years.

In the right situation, SAMP has proven to be an effective means for tackling complex, multidimensional issues within a distinct geo-

Box 2.5 continued

graphic area, such as a watershed, harbor, or littoral cell. To make the process work, it is important to have a broad policy base (e.g., a state coastal management program and other laws), an institutional framework within which implementation can proceed, and recognition among all parties that bargaining is the best way to achieve individual goals.

SAMP is a classic example of a successfully transferred innovation. There are several reasons for its success. First, the timing was right—coastal managers were searching for a new approach to solving the complex problems they faced. Second, and perhaps most important, the process was well documented and fully described. The Grays Harbor planning story was presented at national conferences, written about in coastal management and planning journals, and was a centerpiece in a report to the president detailing progress in coastal management. Finally, key individuals were involved in the process. Especially important were federal agency participants who early on saw its potential for application elsewhere.

FUTURE INFORMATION SHARING—OVERCOMING BARRIERS AND CONSTRAINTS

The interviewees and workshop participants said that many of the methods used today for dispersing data, information, and best coastal management practices—conferences, targeted workshops, and networking generally—need to be continued. At the same time, they noted that the workload and budget constraints that limit travel opportunities make technologies like

Table 2.4   Future Information Sources and Delivery—Obtaining Coastal Management Information in the Future

Variety of sources will continue to be important, including
  Conferences and targeted workshops.
  Interactive, distributed video/teleconferencing that works.
  Reliable "consulting" service that include key contacts, seminal papers and case examples, collective lessons learned, "just-in-time" service.
  Internet resources customized to offer data, information, personally controlled analyses, and issue-focused syntheses.
  Information networks and partnerships that purvey specialized information.

Webcasting and interactive video conferencing increasingly important for information sharing. They also said that it would be valuable to have a reliable, national "consulting" service available as a source for lists of key experts, important papers, case studies, and lessons-learned documentation. Such a service could be provided via the Web, but a human interface was considered important, too. Table 2.4 identifies ways coastal managers would like to get information in the future.

Coastal managers also identified significant obstacles to effective networking and utilization of new and emerging technologies:

- State and federal revenue shortages and the demand to do more (or better) with less
- Workload- and budget-driven limitations on out-of-state travel for conferences, meetings, and workshops
- Shifting social priorities and resource allocations, such as the rising concern for homeland security and terrorism
- Lack of management experience with new technologies and limited local expertise to acquire such experience
- Difficulties in sorting through abundantly available information, much of it useless, to find what they need
- Overwhelming day-to-day pressures from rapid growth and development in the coastal zone
- Recognition that many coastal problems originate "upstream" of the coasts and are beyond the specific scope and authority of many coastal programs

The survey and workshop revealed that reliance upon and familiarity with the products, tools, and services of the many organizations that support and assist coastal managers is somewhat compartmentalized—the stovepipe analogy made earlier. Despite some attempts to overcome this barrier at the federal level—NOAA and the USEPA initiatives on non–point source pollution, for example—it remains a significant problem, particularly between sectors.

Nonetheless, there are potential solutions to these and other constraints, some technological, others institutional. For example, NOAA and the USEPA might undertake a joint strategic assessment of what information is needed and how, together, the agencies might best provide it. Such an effort could improve service to the coastal management community by eliminating duplication, filling gaps, and maximizing the return on resources of both agencies. Jointly developing a customer service plan could also clarify the areas of greatest interest so that these could receive appropriate attention from both agencies. Similar examples exist for other agencies and programs.

CONCLUSIONS AND RECOMMENDATIONS

Over the past three or four decades, the relatively small community of professionals we identify as "coastal managers" has been highly resourceful, innovative, and sharing of ideas and approaches for accommodating people and their demands on the coast while preserving and protecting natural resources. Today, however, coastal lands, rivers, and near-shore oceans and resources are under unprecedented pressure to provide sustained value and benefits to more and more residents, visitors, businesses, and industries. More than half of all Americans reside along our coasts— less than a fifth of our land area—and the rest come to visit. Evidence abounds that coastal and ocean ecosystems are on the verge of collapse (Pew Oceans Commission 2003; U.S. Commission on Ocean Policy 2004). The demand for creative, innovative approaches and practices to manage the people and resources of our coasts and oceans has never been greater. The need to effectively share and adapt the best of those practices is clear.

Many factors drive innovation in coastal management practice. Problems related to the social, economic, and demographic changes alluded to in this chapter are drivers—coastal managers are constantly faced with new challenges demanding new approaches. Policy initiatives are also an important driver of innovation, in part because those changes often come with new resources and authority. Increased scientific understanding of problems, advances in communication and information technologies (discussed further in Chapter 4), and public and interest group demands for more involvement in decision making are other important factors contributing to innovation.

Coastal managers learn about and share innovations through a dense and complex array of networks, the nature of which are explored in Chapter 3. Despite the growing importance of technology in expanding and transforming coastal management networks, personal, face-to-face contact is still considered most valuable—sharing and learning at conferences and workshops were examples often cited in this study's interviews and workshop. Nevertheless, coastal managers acknowledge that to have any hope of success in addressing the wide array of challenges they face, a better way to integrate new communication and information technologies into their practice is needed. At the same time, numerous barriers and constraints need to be overcome, including information overload, limited human resources, increased workload demands, limited technical capacity, and a culture and leadership that often discourage risk taking. Probably the most common frustration reported by coastal managers was not being able to find the information they needed about how others dealt with problems similar to theirs.

Several recommendations flow from these findings. There is a need to strengthen existing coastal management learning networks and build new networks. As discussed in detail in the next two chapters, using and building networks, in turn, will depend on three major types of activities: the shaping of networks, linking coastal management organizations to networks, and using technology to maximum advantage. There is a need to establish an all-purpose information source for innovative best practices, available on demand, when and where a coastal manager needs it. The committee recommends that NOAA establish a compendium of peer-reviewed case studies and examples of innovative or successful coastal management practices—best coastal and ocean management practices, or BCOMPS. The NOAA Coastal Services Center might serve as a national node or coordinator for such a system. Sea Grant's university-based programs in every coastal state could also serve as nodes in such a system. Present technology makes this possible, but there is also a need to link this kind of information to real people—experts who can help new adopters assess their needs, evaluate options, and adapt ideas in real time to meet their needs.

3

LEARNING FROM
LEARNING NETWORKS

ONE CLEAR MESSAGE from the conversations with coastal managers during this study is the importance they give to the networks in which they participate; nevertheless, they also were clear about the frustration they feel about how complex and time consuming it is to be effective members of a network in a world in which the day-to-day demands of the job are often overwhelming. This chapter explores the types of networks in which coastal managers participate and the ways those networks function to facilitate learning and innovation. Also examined are the problems a coastal management organization faces as it strives to be effective in a variety of networks and to adapt innovations from elsewhere to its own circumstances.

SHAPING COASTAL MANAGEMENT NETWORKS

The number of formal and informal networks of people involved in coastal management is beyond counting. Even if the time and energy to create such a count were available, the number would surely be obsolete by the time the process was over because new networks come into existence in the time it takes to copy an e-mail or post something to a Web site. But all of the networks can be said to fall into one of four major types: issue networks, professional networks, sectoral networks, and political networks.

ISSUE NETWORKS

Issue networks are organized around what people do, or the problem or opportunity they are addressing. A good example of an issue network in coastal management is the national Restore America's Estuaries organization, which has grown sufficiently large and formal that it has its own biannual conference. State counterpart organizations have also developed, such as Restore Washington's Estuaries. Another is Sea Grant's hazards network or HAZNET (http://www.haznet.org/), comprised of Sea Grant researchers, outreach staff, and others inside and outside the Sea Grant organization. Similar networks exist for almost any coastal management issue you can name—water quality monitoring, wetland restoration, invasive species control, aquaculture development, and so on. They range from local in scope to national and even international, and may include members from public, private, nonprofit, and other organizations. (Boxes 3.1 and 3.2).

Box 3.1 Beneficial Use of Dredged Materials

About 400 million cubic yards of sediment are dredged each year in the United States by federal, state, and local governments, and private interests such as marinas, in order to keep the nation's waterways open to boat traffic for recreation, commerce, and defense. Disposal of this material in ways that do not degrade the environment is an issue around which informal issue networks have been established around the country. One such example is the Beneficial Uses Group (BUG) organized in 1992 as part of the Houston–Galveston Navigation Channels Project. BUG developed a plan to use dredged spoil to build wildlife habitats and construct 4,250 acres of salt marsh. The plan was designed to partially restore lost wetlands, construct a six-acre bird nesting island, create an underwater berm for topographic relief and fish habitat, and restore Goat Island in Buffalo Bayou. The Port of Houston Authority and the U.S. Army Corps of Engineers (with support from the U.S. Environmental Protection Agency, the U.S. Fish and Wildlife Service, the National Marine Fisheries Service, the U.S. Natural Resources Conservation Service, the Texas Parks and Wildlife Department, and the Texas General Land Office) built levees to allow the construction of more than 900 acres of marshes in Upper Galveston Bay near Atkinson Island using dredged

Box 3.1 continued

materials. The first of these marshes is being completed and evaluated now.

The Houston–Galveston Navigation Channels Project was initiated to better accommodate the variety of vessels in one of the nation's largest ports. The Port of Houston ranks first in the United States in foreign waterborne commerce and second in total tonnage, brings in $10.9 billion annually to the region's economy, and creates more than 287,454 jobs in Texas and another 714,000 jobs nationwide. When this project's feasibility report and environmental impact statement were prepared, one of the main issues raised was where to put the material that was dredged. A part of the U.S. Army Corps of Engineers efforts to solve problems associated with this project was the creation of an Interagency Coordination Team that established the BUG.

The BUG solicited community input to determine how to create the best plan that would enhance both the environment and the surrounding communities. The plan to build wildlife habitats with the dredged materials was presented to the Interagency Coordination Team and received unanimous support from all the federal and state resource agencies. The Port of Houston and the U.S. Army Corps of Engineers share in the cost of construction.

One part of the plan involved the creation of four cells in Upper Galveston Bay near Atkinson Island. One cell was filled in spring 2002 and the remaining three cells will be filled with material dredged for ship channel maintenance over the next twenty years. Additional cells will be constructed to provide capacity for the maintenance material that will be dredged over the fifty-year life of the project.

Once a cell is filled, it takes about two years for the water to drain out of the dredged material and the material to settle to intertidal level. By the time the material reaches intertidal level, creeks and ponds will have naturally formed or have been constructed. Vegetation will then be planted on the "new" wetland cell. Unused cells will serve as open-water lagoon habitat until they are filled.

The BUG maintains a Web site at www.betterbay.org, which is linked to the Port of Houston Authority's Web site, among others. The Army Corp of Engineers tries to communicate widely to spread information about the BUG. The group is in the process of publishing a report on the initial marsh project that will detail almost forty lessons learned on such topics as achieving elevation, providing tidal exchange and circulation, establishing vegetation, monitoring, interagency and public coordination, and construction contracts. These lessons should be useful to others interested in the beneficial use of dredged material.

Box 3.2   Ducks Unlimited: Private Lands Stewardship

The Ducks Unlimited, Inc., private lands stewardship program is a good example of a growing network built around the issue of habitat restoration. Ducks Unlimited biologists work with private landowners, many of them farmers, in a nonregulatory manner that promotes stewardship of private lands and restoration of wetlands. This effort has helped farmers reduce the use of marginal farmland and restore riparian and wetland habitats.

Because over 70 percent of the remaining wetlands in North America are privately owned, Ducks Unlimited recognized that conservation work on public lands alone would not suffice. Indeed, if wildlife is to have a stable future in the United States, wildlife habitat must be secured in private lands. The challenge was to find the incentives (such as cost sharing and technical assistance) that made it possible for private landowners to integrate wildlife habitat into their farming operation.

Ducks Unlimited championed the private lands approach in the Chesapeake Bay Watershed. Since 1998, Ducks Unlimited, along with other partners, has created and restored more than 50,000 acres of wetlands and riparian buffers throughout the Chesapeake Watershed. Bay scientists have estimated that this effort in the state of Maryland has helped to keep more than 6 million pounds of nitrogen, 520,000 pounds of phosphorous, and 188 million pounds of sediment from entering local streams and ultimately the Chesapeake Bay.

Most notably, the Chesapeake Bay Foundation has adopted the Ducks Unlimited private lands habitat restoration model. Many state fish and wildlife agencies and federal agencies conduct private lands habitat work. The unique attribute of the Ducks Unlimited approach, however, is that it is nonregulatory and voluntary. As a nongovernmental organization, Ducks Unlimited is better positioned than other organizations to establish a rapport with landowners to do this type of restoration work on their private property.

The Delaware Coastal Management Program (DCMP) has also adopted the Ducks Unlimited approach, after becoming familiar with it through projects completed on Maryland's Eastern Shore. The land use on the Eastern Shore closely resembles the private lands opportunities that are available in Delaware. On the Eastern Shore, Ducks Unlimited biologists have been responsible for putting some 1,100 projects on the ground. Their work with private landowners has resulted in over 12,900 acres of wetlands, riparian, and upland habitat restoration. The DCMP, by entering into a private land partnership with Ducks Unlimited, is able to engage private landowners in a nonregulatory and non-adversarial way to accomplish several of

Box 3.2 continued

the program's primary objectives. Furthermore, the DCMP is able to tailor this approach and focus the partnership's efforts in the Blackbird and St. Jones watersheds. These watersheds contain the program's National Estuarine Research Reserve holdings.

Demand from landowners for this type of program outpaces current capacity. Of the 3,500 projects that Ducks Unlimited has completed in the Chesapeake Bay watershed, only three landowners have ever opted out of their fifteen-year restoration agreements and taken their land out of habitat management.

This approach is discussed further and illustrated by photographs on the Ducks Unlimited Web page at http://www.ducks.org/conservation/projects/GreatLakesAtlantic/index.asp. For further information contact Mr. David Carter, Program Manager, Delaware Coastal Management Program at 302-739-3451.

PROFESSIONAL NETWORKS

Engineers, scientists, lawyers, economists, planners, and virtually all professions have some form of organized core network of people working in the field, and there are often many subnetworks of people who pay particular attention to specific subjects. Subnetworks are sometimes organized in parallel with issue networks, as is the case for coastal planners. Professional networks, however, are vitally concerned with the state of the profession (education, ethics, standards, etc.), as well as with the applications to specific issues. More important, professional networks are organized around how people are trained to examine issues.

SECTORAL NETWORKS

Sectoral networks form around where people work or the particular economic or governmental subdivision of which they are a part. There are sectoral networks for federal, state, and local government officials, for those working in nongovernmental organizations, in the business community, and in various industries—marine transportation or the fishing industry, for example. Sectoral networks may combine features of the issue or professional network, but these networks primarily link those with common concerns in the same sector (Box 3.3).

Box 3.3 The Green Hotel Movement

A good example of a sectoral network of great importance to coastal environmental health is the green hotel movement. Green hotels are those that have instituted programs to save water, conserve energy, and reduce solid waste to protect the environment. Many are located along America's ecologically sensitive coasts. These hotels have instituted many small operational and managerial changes that have saved them thousands of dollars, improved their operational efficiency, and reduced the environmental damage they inflict on the coast.

Hotels and resorts can produce large quantities of wastewater from common practices that guests expect: graywater, which comes from washing machines, sinks, showers, baths, and roof runoff; and blackwater, which comes from kitchen dishwashing and toilets. Often, wastewater spills untreated into the environment, carrying pollutants such as fecal coliform bacteria and chemicals into the ground or surface water. This can lead to the degradation of particularly sensitive marine environments, including coral reefs and beachfronts. Failure to treat this waste properly has resulted in human illness, including infections, gastrointestinal diseases, leptospirosis, and cholera, as well as environmental contamination, including beach closures (Sweeting and Sweeting 2003).

Hotels of all shapes and sizes across the United States, ranging from small inns to Disney World, are taking measures to conserve, reuse, and clean up the water that they use, so as prevent damage to their guests, neighbors, and surroundings. One simple practice many guests notice are the signs in hotel rooms asking guests to consider using their towels more than once, reducing the amount of water used in washing linens daily. As reported by Sweeting and Sweeting (2003), some hotels are doing more to reuse water. For example, the Apple Farm Inn and Restaurant in San Luis Obispo, California, started using discharged water from washing machines to flush toilets, saving 4,200 gallons of water a day and $5,000 a year. On a larger scale, and equally cost effective, instead of using municipally treated water to irrigate landscaping and golf courses, Walt Disney World Resorts in Florida started using four million gallons of its own wastewater for that purpose. These are just a few of the hundreds of practices being instituted in the United States and around the world.

Making hotels and resorts more environmentally friendly is not necessarily expensive or complicated. Installing devices that reduce and reuse water is often cost efficient; even if there is an initial outlay of funds, the returns can be large—the hotel saves money while publicizing the fact that it is helping to protect the environment. The practices of green hotels can be a win–win situation for the hotel

Box 3.3 continued
industry, the coastal regions, the hotel guests, and the local communities alike.
   Green hotel ideas have diffused rapidly as ecotourism booms
across the country. There is also expanding research about the
affordability and practicality of developing environmentally friendly
practices at resorts. The Internet seems to help this diffusion, as several states maintain Web sites with links to information about how to
develop green hotel programs. In addition, www.greenhotels. com is
an online association that assists hotels looking to develop green
practices.

POLITICAL NETWORKS

Coastal management is primarily, though not exclusively, about affecting
public policy at multiple levels of government. Political networks are the
web of relationships that exist to set the agenda for coastal management
policy, to move the agenda toward decisions, to implement those decisions, and to assess their outcomes for the next iteration of policy. Political
networks are the connections that bring people together to get things
done and make them successful.

PARTICIPATION IN NETWORKS

Everyone involved in coastal management is likely to be involved in one
or more issue, professional, sectoral, or political network, and nearly all
coastal management organizations will be involved somehow in three or
four different types. In fact, coastal management organizations and efforts
that seek to be innovative must be effectively connected to all four types
of networks. New ideas can emerge in any of the networks at any time,
and while no person or organization can be fully successful in tracking all
the ideas that emerge, innovative organizations simply cannot afford to
voluntarily leave themselves out of any of the networks.
   Nonetheless, the existence of these networks, each of which plays
a somewhat different role in the process of innovation, does not say
anything about how the networks operate or how they can be made to

operate better to serve the innovation or information needs of coastal management. To do that, networks must be capable of doing more than simply exchanging e-mails on a listserve or holding an annual meeting. They must be capable of encouraging and sustaining learning.

LEARNING IN NETWORKS

As is the case with most specialists, coastal zone managers and other coastal professionals face problems for which they have no answer—or the current answer is deemed to be no longer work effectively. The more energetic seek answers from colleagues, read reports, and search Internet sites and familiar repositories of knowledge. When they find the management technique, database, innovative strategy, or "best practice" that seems to fit, they then adapt it to their problem. This is "learning on demand."

> [Learning is] usually treated as a supply-side matter, thought to follow teaching, training or information delivery. But learning is much more demand driven. People learn in response to need. When people cannot see the need for what's being taught, they ignore it, reject it, or fail to assimilate it in any meaningful way. Conversely, when they have a need, then, if the resources for learning are available, people can learn effectively and quickly. (Brown and Duguid 2000)

While the Internet and other communication technologies have certainly made it easier to learn "effectively and quickly," our increased understanding of how information needs are defined, how information is sought, and how it is applied to real organizational needs reveals a more complex learning-on-demand process in which the possibility of error is surprisingly high. Several characteristics of the organizational search for information to improve practice complicate effective learning:

- *Knowledge is dynamic.* Much of our knowledge about key natural and organizational phenomena is constantly being shaped and reshaped. Our understanding of the costs and benefits of environmental management, conditions leading to algal blooms, optimal conditions for interagency coordination, and best practices for managing non–point sources of pollution keeps changing. Effective

learning on demand requires knowing what to look for and
where the latest knowledge is likely to be located.

- *Knowledge about is different from knowledge how.* Not only knowing that the impacts of new coastal resorts on community infrastructure are cumulative, but also being able to calculate those
impacts with precision, distinguishes knowing about from knowing how. Effective management requires the ability to apply general knowledge about the functioning of ecological systems or
theories of microeconomic behavior to specific times, places, and
conditions. In so doing, it is possible to illuminate those conditions in ways that make positive intervention possible.

- *Value questions may be mistaken for information questions.* The
organization's search for new knowledge may be hampered by
mistaking value questions for empirical questions—or by pur-
posely converting policy questions into technical ones. For example, in urbanized coastal areas with rapidly eroding shorelines,
coastal agencies may be under pressure to approve private construction of seawalls or to actually develop revetments or other
erosion protection structures as a means to reduce erosion rates
and "save" private property. Policy debates—and information
searches—frequently focus on questions related to which erosion
control strategies have proved most effective for particular types
of coastal erosion. These debates about erosion control strategies
may mask more value-laden questions about whether government intervention is good public policy and, if intervention is
undertaken, how costs should be allocated between all taxpayers
and the individual landowners who directly benefit from what
amounts to a government subsidy. Allowing policy questions to
be converted into technical questions can skew a search process.

- *Knowledge is distributed.* Knowledge is sometimes thought of as
residing in a few key experts. Our predisposition to think of
knowledge as an individual phenomenon is partly a product
of our lengthy school experience where we are encouraged to
"learn on our own" and "do our own work." These norms
encourage thinking of learning as a solitary, independent activity
(Dixon 2000). Many of us recognize, however, that knowledge is
frequently constructed in groups.

> Looking at learning as a demand-driven, identify-forming, social
> act, it's possible to see how learning binds us together. People with

similar practices and similar resources develop similar identities—
the identity of a technician, a chemist, a lepidopterist, a train spot-
ter, an enologist, an archivist, a parking lot attendant, a business
historian, a model bus enthusiast, a real estate developer or a cancer
sufferer. These practices in common (hobbies and illnesses are practices, too) allow people to form social networks along which
knowledge about practice can both travel rapidly and be assimilated readily. (Brown and Duguid 2000)

TYPES AND CHARACTERISTICS OF LEARNING NETWORKS

If we recognize that much practice-related knowledge is distributed among
coastal managers, particularly in professional, issue, sectoral, and political
networks, we still have to ask, How can managers define information needs
more effectively, how can they use knowledge networks more effectively,
and how can they apply this knowledge to the issues they confront? To
find answers, the committee examined how the networks that coastal
managers participate in can be effectively transformed into different types
of learning networks. Several key questions are addressed:

- What is meant by learning networks?
- What are the attributes of learning networks?
- What types of learning networks can be distinguished?
- What are the implications of these examples for strengthening
  coastal management learning networks?

All organizations have informal networks of people who communicate, share information, and build relationships and reputations (Wenger,
McDermott, and Snyder 2002). What makes learning networks distinct is
their specific substantive focus (e.g., coastal erosion, marine protected area
management, or hazard mitigation) and the degree to which learning is an
explicit purpose for maintaining the network. Such learning networks seek
to share information, identify lessons from management experience that
can improve practice, test supposed best practices, or seek better understanding of some natural coastal phenomenon. Not all learning networks
are alike. They can be distinguished in terms of a few key characteristics:

- *Boundaries*. What is the substantive focus of the learning network? How well defined is it?
- *Membership*. How open is participation in the network? What
  requirements govern participation? Is access to the network open

to all or only a select few? Are there categories of membership? How, if at all, does the volume of participation matter?

- *Participation requirements.* What expectations do participants in the network have of each other? Is the network perceived to be a "knowledge bank" from which participants can make withdrawals? Are participants expected to contribute knowledge (in the form of research, best practices, or other products) to the bank?
- *Knowledge generation.* How is the knowledge shared among network participants generated? Is knowledge generation supported in some formal way through contracts, grants, or other incentives? What mechanisms, such as peer review, are employed for quality control?
- *Communication technologies.* How do participants in the network communicate with each other? What combination of Internet, telephone, scheduled meetings, conferences and workshops connects them? What modes of communication are deemed most important and valuable?
- *Practice orientation.* How important is the application of knowledge to participants? How important is "knowing how" to apply new knowledge?
- *Network governance.* Is there a central hub or coordinator for the network? Is it self-organizing? How are Web pages and listserves maintained? How tightly managed is the network?

Learning structures, including networks, can be differentiated in terms of several of these attributes (Table 3.1).

Formal Departments and Project Teams

Staff in some departments and, more frequently, on project teams, often exchange information and advice about how to improve service provision or product delivery. The reliance on face-to-face information sharing sets such departments and project teams apart from the learning networks described below. Examples of work teams or project teams that function as learning networks may be found in Wenger (1999), Wenger, McDermott, and Snyder (2002), and Brown and Duguid (2000). Many such examples exist within coastal management organizations and may even be the principal structure for accomplishing objectives.

Table 3.1 Distinctions among Learning Networks

Type of Learning Structure	Purpose	Membership	Boundaries	Network Governance
Formal departments	To deliver a product or service	Everyone who reports to the department administrator	Formal	Administrator
Project teams	To accomplish a specified task	People who have a direct role in accomplishing the task	Formal	Team leader
Professional, political, issue, and sectoral networks	To receive and pass on information; to identify, explore, and advocate particular issues or positions	Members of a profession, interest group, or bureaucratic group	Formal ↔ Informal	Executive committee or other centralized coordinator
Collaborative learning networks	To explore well-defined themes, undertake research, and generate knowledge	Participation by invitation	Formal ↔ Informal	Research coordinator
Communities of practice	To explore well-defined themes, undertake research, and generate knowledge; to apply knowledge to accelerate application of knowledge	Participation by invitation	Formal ↔ Informal	Coordinator or principal investigator

Source: Adapted from Wenger, McDermott, and Snyder 2002.

Professional, Political, Issue, and Sectoral Networks

Informal professional, political, issue, and sectoral networks exist to share knowledge and information among what may be a broadly distributed network, such as coastal managers, federal environmental officials, or commercial fishers. Participants may be members of a professional association in which membership is open to any who pay dues. The American Planning Association (APA) can be thought of as just such an informal learning network (Box 3.4).

A good example of a professional network within the coastal management community is The Coastal Society (TCS). TCS was founded in 1975 as an international professional organization for coastal managers—planners, scientists, engineers, attorneys, and others who had migrated to this new field. Its 400 members represent the core of the U.S. coastal

---

Box 3.4   The American Planning Association

The mission of the American Planning Association (APA) is to provide "leadership in the development of vital communities by advocating excellence in community planning, promoting education and citizen empowerment, and providing the tools and support necessary to effect positive change." The substantive domain of the APA is quite broad, as reflected in its eighteen divisions. There are no participation requirements other than paying dues. APA members can be as active—or passive—as they like.

The APA publishes the *Journal of the American Planning Association* and operates a Web site (www.planning.org) that offers links to information on publications, conferences and workshops, jobs and careers, and pending legislation affecting planning. Knowledge generation is the responsibility of paid consultants, members who publish books and monographs, and a seventeen-person research staff that organizes and conducts applied research. The Planning Advisory Service is well known to planners. Reports are commissioned on a wide variety of topics of interest to some planning constituency.

The APA operates as a specialized, interactive library. It offers information and knowledge thought to be of interest to significant constituencies within the profession. It organizes national and regional conferences with presentations and panel discussions on a variety of general and specific topics. The breadth of the field and the openness of membership make the APA a loose, informal learning network.

management community—state and federal coastal zone management leaders and staff, specialists in coastal water quality and protected areas, academics and policy students in the marine affairs programs they serve, and nongovernmental organizations and private-sector specialists who interact with coastal managers in their work. TCS produces a newsletter, holds an international conference every two years, supports regional and student chapters, and provides job announcements and other information to its membership. It is not as large or sophisticated as the APA (Box 3.4), but serves many of the same functions.

The Coastal States Organization (CSO) operates as both an issue and a political network. CSO membership is open to representatives of the governors of the thirty-five coastal states, territories, and commonwealth. CSO "encourages cooperation among the states to resolve national coastal issues without interfering in the states' pursuit of individual, and sometimes, differing objectives" (http://www.sso.org/cso/aboutcso.htm). CSO advocates for state coastal managers in Congress and in the federal bureaucracy, coordinates legislative proposals, and disseminates information to members' states, among other things. Communication among members of the network is primarily via phone, e-mail, and meetings several times a year. CSO is administered by an executive director and directed by an executive committee whose members represent a cross-section of states in each region.

Nonprofit organizations are increasingly recognizing the benefits of sectoral networking in order to maximize the benefits of their individual and shared efforts, to avoid duplication of effort, and to stretch the successes from limited resources. Often these networks involve limited geographical focus, a particular project focus, or a limited subset of the environmental community that share a common vision or strategy. For example, the fast-growing Waterkeeper Alliance is a closed membership of the one hundred or more River-, Sound-, Bay-, and Inlet-keepers programs emerging across the country. Although each of the individual programs is a distinct organization with its own nonprofit status, governing board, financial resources, and watershed of focus, the national Waterkeeper Alliance was created to encourage information sharing and collaboration on common issues (http://keeper.org).

The National Oceanic and Atmospheric Administration (NOAA) Coastal Services Center (CSC), described earlier in Box 2.4 (pages 39–42), sometimes operates as an informal sectoral network, disseminating technical information among coastal managers. CSC's geospatial technology workshops and training sessions on geographic information systems, meta-data (data about data), and other topics are good examples of this role as information provider (http://www.csc.noaa.gov/bins/training.html). At other times, however, the CSC functions more as a formal collaborative learning network in which it partners with specific organizations to analyze some coastal issue or create some new management process. The CSC's partnership with New Hanover County, North Carolina, to develop a Community Vulnerability Assessment Tool illustrates this role (http://www.csc.noaa.gov/products/nchaz/startup.htm).

Collaborative Learning Networks

Collaborative learning networks are formal knowledge networks that create and disseminate knowledge for use beyond the membership of the network. According to Clark (1998), such knowledge networks have several key characteristics:

- Their structure and operation are designed to maximize the rate of knowledge creation.
- The network must provide recognizable direct benefits to all participants.
- There is a formal organization and well-defined management structure.
- Participation is by invitation, based on criteria of merit or peer review.
- There is a well-developed communication strategy.

One example of a collaborative learning network is given in Box 3.5. Another is the Ocean Governance Study Group, created in 1991. More than thirty academics with strong interests in ocean and coastal management participate in the network. Members of the network identify key governance issues in ocean and coastal management, share research and policy papers, prepare testimony, and participate in periodic workshops. The Center for the Study of Marine Policy at the University of Delaware serves as the network coordinator.

Communities of Practice

Communities of practice are another example of formal knowledge networks. The primary distinction between collaborative learning networks

Box 3.5   The Packard Foundation's Integrated Coastal Management Sustainability Project

The Packard Foundation's Integrated Coastal Management Sustainability Project was established to examine factors affecting the success and long-term viability of community-level coastal management projects in the Philippines and Indonesia. As a collaborative learning network, the project connects scholars from several universities in the United States, Indonesia, and the Philippines, along with nongovernmental organization staff. These participants share related research interests or field experiences. U.S. scholars are working with Philippine and Indonesian counterparts on empirical research into coastal conditions, socioeconomic factors affecting local success, institutional capacity, and intergovernmental relations.

Participants in the network conduct research, write papers, and exchange papers with participants in the project and other colleagues. Project team members have had three workshops to present and discuss their research, to refine the questions they are addressing, and to discuss research methods. While the primary purposes of the research are to explain the dynamics of community-level coastal management, the work has obvious implications for the practice of community management broadly. Refining local management strategies is not, however, a central purpose of the project.

Governance of the network is shared, but the principal investigator bears primary responsibility for keeping the Web site up to date (http://www2.mozcom.com/~icm_proj/), managing the network, organizing meetings, maintaining relationships among the network participants, and funding.

and communities of practice is that the latter conducts research primarily to improve practice. A good example of a formal knowledge network that meets our criteria as a community of practice is the Locally Managed Marine Area (LMMA) network (Box 3.6).

At the national level, Sea Grant has organized communities of practice around themes, several of which are strongly linked to coastal research and management—coastal hazards, coastal communities and economies, the "digital ocean," fisheries, ecosystems and habitats, and the urban coast. These teams are comprised of Sea Grant experts from more than thirty universities along the nation's coasts and serve a variety of learning network functions: initiating joint ventures; directing funds to high-impact areas; providing a forum to organize research, extension,

Box 3.6   The Locally Managed Marine Area Network

The Locally Managed Marine Area (LMMA) network is a good example of a growing community of practice. The LMMA network is a system of community-level marine management projects (referred to as a "learning portfolio"*) in the Pacific organized to achieve three goals: implement effective conservation projects, learn about the conditions under which this conservation strategy works or does not work, and improve the capacity of the members of the portfolio to do adaptive management.

A learning portfolio is defined by projects that are using a common conservation strategy. The basic unit in a learning portfolio is a project. Projects typically try to achieve conservation at one or more specific sites.

Membership in the portfolio is by invitation. Portfolio members are categorized as:

- *Full members*: Communities, organizations, and projects that are currently implementing locally managed marine areas that have completed initial membership obligations (i.e., obtaining support from local partners, completing an initial site description, developing a monitoring plan, collecting baseline data, and appointing project representatives to the portfolio).
- *Provisional members*: Sites and communities that are interested in joining the network, but have not yet completed initial membership obligations.
- *Network members*: Sites, communities, and projects that do not want to be fully involved in the portfolio, but want to be part of a broader network of practitioners and researchers focusing on the subject.
- *Donor members*: Funders supporting the portfolio or work at particular sites. They can be decision-making members of the portfolio with the approval of two-thirds of the full membership (http://www.lmmanetwork.org/; Nickerson and Olsen 2003).

At present, the LMMA has twenty full members in eight countries. Ten projects are in the Pacific: Fiji, Papua New Guinea, the Solomon Islands, the Cook Islands, and Palau; and ten projects are in Southeast Asia: Indonesia, the Philippines, and Malaysia. There are currently twelve provisional members (Nickerson and Olsen 2003). Access to technical assistance, a shared conservation philosophy, a community of project practitioners, and travel funds are among the incentives to participate in the LMMA network.

---

* For more information on learning portfolios, see (www.fosonline.org).

Box 3.6 continued

At the center of the LMMA network is the Network Coordination Team (NCT), composed of six to eight individuals from the LMMA projects. NCT members and project staff coordinate the overall activities of the network by "planning the network, diagnosing individual project needs, and working with them on a regular basis, coordinating cross-project activities, coordinating portfolio level analyses, and helping to communicate results" (Nickerson and Olsen 2003).

The LMMA network has organized a variety of learning activities, including workshops, cross-project visits, training sessions, and portfolio meetings. The network also provides technical assistance, supports study tours, promotes sharing logistical and technical information among members, produces "stories" showing project successes and failures, produces educational materials and policy briefs, and fosters linkages with resource people at key institutions and communities (Nickerson and Olsen 2003). The network has also Web site (http://www.lmmanetwork.org) to store data, analytical results, and reports.

communications, and education efforts on regional and national scales; and facilitating information transfer to coastal managers and policymakers (Tracy Crago, personal communication, June 27, 2003).

IMPLICATIONS FOR COASTAL MANAGEMENT

Coastal managers are not passive participants in their networks, merely waiting for the next e-mail or meeting. All the networks in which coastal managers participate are in one way or another owned and managed by the participants, whether through a representative institution such as a board of directors or through the initiative of individuals. As owners, the participants take responsibility for making the networks function as well as possible, so coastal managers need to be continually involved in deciding how each of the networks in which they participate can be improved.

Of course, this commitment to improvement does not imply that every network must be transformed into a highly organized, highly formal community of practice. Such a goal would surely be enormously wasteful of resources in many areas where nothing more than a listserve or a lunch at the annual meeting is sufficient to convey the current state of knowledge. Nonetheless, each participant must be involved in a continuous process of self-assessment about the network's strengths and weaknesses as a learning network. Participants must also recognize and be prepared to take action when the time comes to change the form of the network. Participants need to assess how the network contributes to their practice and how they could get more from the networks to which they are linked. They need to ask what communication strategies and what governance arrangements would serve them more effectively. They also need to examine what barriers in their own organizations inhibit more effective learning from their networks.

LINKING ORGANIZATIONS TO THE NETWORKS

A key challenge in making networks more effective is determining how to link individual organizations to larger networks. The process is not automatic, particularly if the network is to be a major source of innovation. Four characteristics of the organization that greatly influence how successful networks will be in fostering and diffusing innovation are organizational culture and leadership, resources, roles, and network connections.

Organizational Culture and Leadership

To what extent does the culture of the organization support learning and innovation? Organizations and their leaders must choose to be innovative and must make learning part of their routine operating approaches. Without this choice, none of the other parts of innovation by design will matter.

The task of changing organizational culture to promote innovation falls ultimately to the leadership of the organization. An entire library of studies of innovation indicates that without signals from the top leadership that innovation is desirable, no organization in any sector will be innovative except by chance. Leadership can also be extremely important in adopting and spreading innovations used elsewhere. It is very easy for organizations seeking to be recognized for their own contributions to shun any ideas that come from somewhere else, or for individuals who perpetually see their own problems and jurisdiction so unique as to discount any experience gained elsewhere. Fortunately, leaders in coastal management have a number of tools to promote a culture of innovation. These include large-scale activities such as strategic planning; mid-scale activities such as annual budgeting, personnel hiring, and reviews; and small-scale activities such as routine staff meetings. All organizations differ in the extent to which these activities matter—how much flexibility is available in budgets, for example, and the extent to which those in leadership encourage creative problem solving and risk taking. But if an organization is to be innovative, appropriate ways must be found to use all of the activities in support of that objective.

Resources

As with all tasks, participating in learning networks and finding ways to be innovative requires time, energy, money, and people. These are inherently limited, and adding "being innovative" to the existing demands on resources always seems like another potentially back-breaking straw. Although many organizations and public agencies in particular are now under severe budget constraints, any organization that decides to be innovative can deploy the resources to make it happen. Resource allocations flow from decisions about what the organization's tasks are, not the other way around. To be sure, it is likely that not all the resources needed will be available, but then it becomes a matter of using existing resources as well as possible and building for the future.

Roles

One of the biggest challenges in participating in learning networks is determining who will play the role of the local "node" on the network. Ideally, everyone who works in an organization should be part of the search for innovative ideas, but it is difficult to simply add this task to everything else that must be done. Sending everyone on staff to a conference or assigning everyone to search the Web every day is costly and inefficient.

Although it is costly, an obvious alternative is to assign someone specifically to be the organization's node on the learning network. Organizations rarely have personnel to spare on such tasks, and even more rarely have the ability to add staff to perform these functions. Often organizations assign these tasks to interns or other temporary employees. This strategy supplies the person-power, but the success of searches for innovation often depends greatly on the experience of the searcher to be able to separate useful from useless information.

Hence, the challenge is to find the best balance between widely dispersing the responsibilities and concentrating them on particular people. There is no universal answer, because every organization is different. The successful balance will most likely involve a combination of specialized roles for some staff, with everyone taking at least some of the responsibilities.

Network Connections

Coastal management already has extensive networks. State coastal management programs, the Coastal States Organization, the Coastal Society, associations of other state agencies (e.g., environmental protection, resource management) all comprise networks with both formal and informal connections. There are Web sites, listserves, annual conferences, and trade and professional associations. The problem is not to create new networks, though that may be desirable in some instances, but to make the best use and even transform the networks that already exist.

One of the keys to making learning networks work effectively as policy tools is to give attention to how connections are made within the organizations participating in the larger networks. A simple diagram shows the problem (Figure 3.1).

The coastal management organization operates within a network of other organizations and individuals, with communications flowing throughout the network. However, the organization itself is depicted as

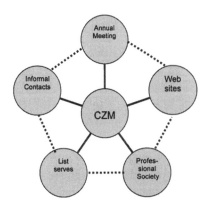

Figure 3.1 How Learning Networks Work

just an undifferentiated circle. How is all that information handled within the organization? How is information from an annual meeting or professional journal transmitted within the organization? If something interesting comes across a listserve, how does it get to the right place in the organization? Is the organization internally "wired" to take advantage of all of its external connections? Various strategies for such wiring, some time-honored and others largely untested, range from routing slips and brown-bag lunches to organizational intranets.

RECOGNIZING AND TRANSFERRING USEFUL INNOVATIONS

Recognizing an idea, program, project, or activity as a potentially useful innovation does not mean that it is applicable in a different setting. An important part of the successful search for innovations is to be able to identify the characteristics that make for transferability as well as utility. Several characteristics have been identified as particularly important (see Rose 1993):

- *Context Dependency.* Programs that address issues that are unique to the area or time in which they are implemented are not likely to be highly transferable. Because coastal environments vary so widely, management successes may depend on their own local context. As a result, innovative programs for water quality management, for example, may be very different in Alaska and Florida.
- *Substitutability of Institutions.* Programs are most transferable when the organizations responsible are similar in scope of authority. For coastal management, however, programs vary greatly in the types of organizations that implement them. In some states, management is centered in an agency with broad responsibilities for coastal-related issues and substantial implementation authority. In other states, it is centered in a policy-oriented agency that must work with line agencies to actually take action.
- *Equivalence of Resources.* Innovations that require many resources are harder to transfer than those that do not, simply because the task of acquiring sufficient resources becomes a barrier in itself. These characteristics make programs like the Coastal Zone Enhancement Grants in the Coastal Zone Management Act particularly valuable because funds are provided to apply to innovative practices, such as special area management planning (Box 3.7).

**Box 3.7   The Pea Patch Island Plan:
Taking SAMP to the Next Level**

The Pea Patch Island Heronry, a Special Area Management Plan (SAMP), was developed through an innovative process that had not been used in Delaware, although upon reflection it seems intuitive. The National Oceanic and Atmospheric Administration (NOAA) sent key personnel to the Delaware Coastal Management Program (DCMP) for several years to train its staff in the intricacies of goal setting, issue characterization, and conflict resolution. The process was so successful that the DCMP used the process for several other controversial projects, including a dredging plan for the state and environmental indicators to protect the coastal zone from industrial development.

Pea Patch Island Heronry is a state and national resource; it is the largest multi-wader heronry north of Florida. The heron population was in decline, and reproductive success was low, largely because of land use changes, associated water quality impacts, and contamination of the herons' prey by pesticides and industrial contaminants. Because potentially responsible resource users were eager to blame each other and reluctant to act, the state needed to follow a fair and transparent process that involved all stakeholders and resulted in implementation of strategies to help the herons.

Senior National Ocean Service staff from the Special Projects Office agreed to help DCMP with the SAMP using a process they had had success with on controversial National Marine Sanctuaries issues. A core group of individuals agreed to follow the process. The process set a goal to protect the heronry, characterized issues that endangered the goal, developed and ranked strategies to tackle the issues, and prepared an implementation plan. At each step, core stakeholders met at a facilitated workshop, and a document was produced to record the workshop's outcome. A separate task group was established to guide needed research (Delaware Coastal Program 2001).

The DCMP has successfully used the same process for other projects; the process has become an established management protocol within the DCMP network. People who worked on the SAMP are aware of the process and have suggested using it for other situations, although diffusion has not extended beyond the existing DCMP network.

Because this process involves a lot of time and planning for meetings, the Pea Patch project would not have been possible without the human resources allocated from NOAA. In the absence of such resource allocations, organizations and individuals outside the network have not been able to utilize the process. Although the process is successful, organizations must be willing and able to change their current approach to problem solving.

- *Complexity of the Program.* Program complexity is defined by the number of connections between causes and effects that programs attempt to address. Simple programs have few connections; complex programs have multiple connections and multiple layers of connections. This simplicity–complexity dimension does not address the innovation of the program, but rather its transferability from one location to another. Although simple programs are easier to transfer than complex ones, complex programs are certainly transferable if strategies can be developed to address the elements of complexity.
- *Scale of Change.* Small changes are easier to make than large changes. This rule applies no matter what the degree of innovation involved.
- *Interdependence.* When the jurisdictions involved in a potential program transfer are interdependent—because they share a resource, face a common scarcity, or are affected by the same problem—it can be easier to make the case for transfer. Put another way, shared problems will make it easier to develop common solutions.
- *Values Congruity.* Policies and programs arise in different political cultures, and those cultures are more tolerant of some approaches than others are. For example, some political cultures value extensive public participation in the program development process, while others do not. Some jurisdictions are more favorable to regulatory programs than others are.

None of these cultural dispositions is an absolute barrier to transferring innovations between jurisdictions, but all influence the degree of difficulty in the transfer. These characteristics can function as a checklist for assessing how easy or hard the transfer of an innovation is likely to be and for developing strategies for dealing with those aspects of the transfer— or the program—that make the innovation hard to adopt.

CONCLUSIONS AND RECOMMENDATIONS

Networks are the key sources of learning about what is new, what works, and what does not. Coastal management is awash in a dense array of networks, most of which function well to sustain some level of innovation, but few of which have been seriously examined by their participants for their ability to be real learning networks. Moreover, coastal managers and organizations often take a haphazard approach to participation, with no overall sense of how each organization can best use its resources as a member of the network. Networks need to be more than the sum of their e-mails and meetings if they are to be true learning and innovation networks.

At the same time, the dramatic development of information technologies has greatly expanded the need and opportunity for new networks and has greatly increased the reach and efficiency with which networks can function. Many see technology as the answer to all of the complications of participating in so many different networks. As we see in the next chapter, the promise of technology to greatly enhance coastal management networks is real and is already being captured to some extent. However, much remains to be done, and the promise of technology should not be overstated. What really matters is still the people sitting at the computers, not the computers themselves.

Much of learning in coastal management, as in other fields, is demand driven, not supply driven. This has significant implications for how organizations with missions focused on technical assistance, outreach, and professional development approach their responsibilities. Information purveyor organizations like Sea Grant, NOAA's Coastal Services Center, and the National Estuarine Research Reserves (see Box 2.4, pages 39–42) understand this and regularly conduct needs assessments to plan their work. However, it is clear that existing networks and programs do not provide the information about innovations and best practices that coastal managers say they need, when and how they want it. Existing networks need to be transformed or new networks invented to provide this information.

Another specific recommendation flowing out of the study of networks as learning tools is the need to increase the flow of people among organizations, not just information. In this regard, we recommend expanded use of the federal interagency personnel act, which provides for short-term personnel exchanges among governmental, nongovernmental, academic, business, and other eligible organizations. The cross-training approach is underutilized in coastal management, yet is an extremely effective way of fostering learning networks and deepening understanding of and appreciation for best practices used by other organizations.

4

The Promise and Limits of Technology

In just the past few decades, a remarkable revolution has occurred in how we collect, organize, and access information—a sea of change brought about by the development of advanced communication, information, and sensor technologies. Moreover, such technologies have enhanced the learning experience, not only by increasing access to information but also by creating new mechanisms for learning and processing information. As a result, individuals and communities are faced with an onslaught of new facts, data, and images, which are almost instantly accessible—with the right tools and knowledge base—from almost anywhere in the inhabited world. This technological transformation in the way information is obtained clearly impacts how we learn, both as individuals and as members of communities. Just as access to information has increased enormously, so have the linkages among learners and communities around the world. But these new opportunities—and new requirements—for learning are not easily understood or embraced.

THE INTERNET AND THE WORLD WIDE WEB

Possibly more than any other invention, the Internet has revolutionized the way people communicate. A worldwide system of computer networks, and networks within networks, allowing communication among hundreds of millions of people, the Internet began in 1969 at the U.S. Defense Department's Advanced Research Projects Agency (ARPA), through which

"ARPANet" was created to allow communication among research computers at different universities (Salus 1995). Such a network was designed to function even if individual computer components were not active, whether because researchers were occupied elsewhere or because equipment was destroyed through military action or some other force. Taking advantage of public telecommunication networks, the Internet has expanded enormously and is now available to virtually any citizen who can purchase a computer or visit a library.

The World Wide Web (WWW, the Web) uses the Internet to move information. It consists of client computers, which receive the information and send commands via software "browsers," and servers that deliver the information. The communication between client and server computers is most often accomplished using the hypertext transfer protocol (HTTP), as well as more dynamic mark-up languages such as Java Script and XML (extensible markup language). This protocol and the related languages specify how messages are formatted and transmitted and how servers and browsers respond to various commands.

Web interfaces have become increasingly sophisticated as the volume and diversity of information they deliver has increased. To provide more efficient access to information, Web portals have been developed to serve as a starting point for clients when they log onto the Web. General portals include sites such as Yahoo, America Online's AOL.com, or Netscape. Other portals, often called niche portals, target a specific interest, such as investors, daily news seekers, hobbyists, etc. Even with such gateways to organize access to information, the amount of information can be daunting. Accordingly, tools for locating specific types of information—search engines—have become one of the most important utilities for working with the Web. Some search engines, such as Google.com, scan across the entire Web to gather references; others are built into portals, so that specific information is easily located within a Web site. Web site designers have recognized that sites can be constructed in ways that improve the probability of their being located and opened by searching users. This has resulted in a tension between the "buyers" and "sellers" of information, who may have different interests, whether or not their interactions involve a transfer of money: as buyers, users often want to quickly locate a few specific bits of information within an extraordinarily large pool of data, and as sellers, providers generally want their products to be found on as many searches as possible. Coastal managers, like every other segment of society that uses the

Web to access information, need improved mechanisms for locating and retrieving data of interest.

GEOGRAPHIC INFORMATION SYSTEMS

Another major technological development, which has created a new dimension for presenting and analyzing ocean and coastal data, is the geographic information system, or GIS (Clarke 2002). GIS is a system that can be used to input, store, retrieve, manipulate, analyze, and output geospatial data or data that are geographically referenced. GIS uses special software and procedures that enable common treatment and visualization of data on separate computers. The sources of the geospatial data can be maps that have been digitized or aerial or satellite images, as well as their associated data. In practice, all GIS data associated with a specific place are geographically referenced to a map projection in a coordinate system, so that it is possible to identify the spatial relations between different features or attributes. These relations can be projected on map displays as layers of information, which graphically illustrate relations and collocation. Such relations can also be quantitatively assessed through statistical analyses, which provide more rigor and help to define the level of confidence in specific purported relations. GIS analyses are often invaluable in identifying the relations between physical features, natural resource characteristics, human activities, and demographic changes. These new capabilities have revolutionized land management and land planning, as it is now possible to search, display, analyze, and model spatial information and to combine both location and attribute data for assessment of management or policy options (Jensen 2000).

A number of programs have incorporated GIS technologies to optimize decision making. One example is North Carolina's Coastal Region Evaluation of Wetland Significance (Box 4.1), which includes wetland site characteristics, water quality classifications, and habitat community characteristics to help identify priorities for wetland protection (Sutter et al. 1999). That model has been further developed for general application by the National Oceanic and Atmospheric Administration (NOAA) Coastal Services Center as SWAMP—Spatial Wetland Assessment for Management and Planning (Sutter 2001).

Another innovative use of GIS technology is the Ocean Planning Information System (OPIS). The system was developed by NOAA in

---

**Box 4.1   Using GIS in North Carolina for Evaluation
          of Coastal Wetlands**

One of the more sophisticated large-area functional assessment systems for existing wetlands is the North Carolina Coastal Region Evaluation of Wetland Significance (NC-CREWS) system. NC-CREWS uses
digital data arrayed in a number of GIS data layers: wetland boundaries, wetland types, soils, land use, land cover, hydrology, watershed
boundaries, endangered species occurrences, estuarine primary nurs-
ery areas, and water quality classifications. The NC-CREWS system
analyzes these data to determine wetland functions across the entire
landscape, generally at the level of individual watersheds. The results
can be used to set priorities for protection of significant wetlands.
The NC-CREWS system has also been adapted to identify and evaluate the restoration potential of former and degraded wetlands.
Using similar GIS techniques, the system can also help to evaluate
qualitatively the potential for sites to perform hydrologic, water
quality, and habitat functions, with scores then combined to give an
overall restoration potential rating. The functional assessment capabilities of NC-CREWS have also been applied in local land use planning and classification mapping, as well as in the federal consistency
review of Clean Water Act Section 404 permits. The NC-CREWS is also
a key analytical and funding component of the recently passed state
legislation on wetland restoration. The NC-CREWS project began
with Coastal Zone Management funding and received additional
funding from the U.S. Environmental Protection Agency.

---

partnership with several Southeast states to include not only georefer-
enced data, but also information on jurisdictional boundaries, relevant
policies, and regulations. OPIS also supports online mapping, allowing
users to create and print maps to their own specifications (Box 4.2). A
third example of the use of GIS to optimize decision making is the development of the Social Vulnerability Index, or SOVI (Box 4.3).

NEW SENSOR TECHNOLOGIES

Technology has not only provided new ways of dealing with information,
but has enormously expanded the volume and kinds of information available to us. In particular, new sensor technologies—evident in the increasing

Box 4.2 Regional GIS as an Information-Sharing Tool for Ocean Management

The Ocean Planning Information System, or OPIS, is a prototype GIS-based planning and marine decision-support system for the southeastern United States, including territorial sea and federal waters offshore of the Carolinas, Georgia, and Florida. Developed by the National Oceanic and Atmospheric Administration (NOAA) Coastal Services Center and NOAA Ocean and Coastal Management, in partnership with southeastern states, the goal of OPIS is to provide easy access to comprehensive ocean-related data and information that will enhance regional, integrated approaches to coastal and ocean resource management. Modeled on Florida's Marine Resource Geographic Information System, OPIS includes georegulatory data for ocean areas in the Southeast, spatially displaying political, legal, and administrative jurisdictional boundaries, along with references and narrative summaries of relevant policy and regulations. Access to OPIS data, online mapping, and other resources are publicly available on the Web (http://www.csc.noaa.gov/opis/).

Both the federal government and states have sovereign powers with separate and sometimes overlapping political, legal, and administrative jurisdictions over offshore waters. Furthermore, within and across governmental levels ocean management is characterized by single-purpose regimes (e.g., fisheries, endangered species, oil and gas, ocean dumping) with few effective mechanisms for harmonizing differences. As a consequence, few understand the complexities of

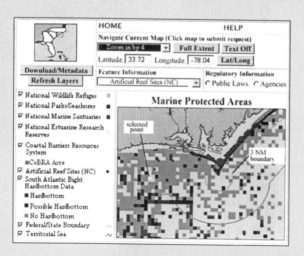

OPIS online mapping application displays bottom types in a marine protected area in North Carolina to assist in determining the suitability of the site for disposal of ocean-dredged material.

Box 4.2 continued

ocean use and jurisdiction, with the result being conflict and often, costly litigation. OPIS, with its spatial display of jurisdictions and supporting policy references, was conceived as a means to ameliorate these kinds of problems. Potential benefits of OPIS (Good and Sowers 1999) include the following:

- The straightforward format and visual nature of the system, as well as its accessibility via the Internet, allow a broad audience of citizens, managers, and scientists to learn more about ocean resources and governance from an ecosystem-scale perspective.
- By mapping jurisdictional complexities to identify gaps, overlaps, and conflicts between laws governing ocean resources and between agencies managing ocean areas, the system improved integration between state and federal government levels and between marine resource agencies operating in the same state.
- OPIS helps to identify conflicts between multiple ocean uses, such as the need to dispose of dredged material or mine sand resources offshore while protecting essential fish habitat and endangered species.

A number of states and regions have established ocean information systems, including the Gulf of Maine (http://woodshole.er.usgs.gov/project-pages/oracle/GoMaine/), Florida (http://www.floridamarine.org/), Oregon (http://www.coastalatlas.net/), and California (http://ceres.ca.gov/ocean/). Each has links to laws and policies, but with no spatial context. An exception is the detailed boundaries in Florida's Marine Resource Geographic Information System Internet Map Server. Even in that system, however, relevant policies and regulations are not linked to the boundaries as in OPIS. OPIS has yet to be emulated in other regions, although recommendations for improved, more integrated, regional ocean management (U.S. Commission on Ocean Policy 2004) may provide impetus for other regions to do so. Lessons learned from OPIS and similar ocean GIS efforts include the following.

- Ocean GIS has a significant initial cost, which is an obstacle to gaining support from government leaders and planning for future maintenance and improvements.
- Ocean GIS still has technical limitations, which make it difficult to integrate and display the kinds of oceanographic and marine ecosystem data that scientists produce (often three- or four-dimensional).
- Coastal managers receive limited technical training on marine science issues; conversely, the science community has limited understanding of ocean management and ocean GIS needs.
- At the state level, ocean management and GIS has low priority for staff time and financial resources, compared with other coastal issues.

Box 4.3   Assessing Vulnerability to Coastal Hazards

The Social Vulnerability Index (SOVI) is a methodology for assessing social vulnerability assessments and creating a statistical index using a consistent set of indicators that facilitate comparisons among diverse places. SOVI enables coastal managers to examine the geographic components of social vulnerability and the temporal variability in the human occupancy of hazardous areas.

What stimulated this innovative approach was recognition of the fact that interactions of humans, the built environment, and biophysical vulnerability all contribute to the overall vulnerability of places. Depending on the focus, these places can range from the smallest unit, such as a home, to a neighborhood, city, county, or beyond. If measurements are made in similar ways, coastal managers can compare the relative vulnerability of various places to see where the burdens (social and structural) and the risks (biophysical) are greatest.

In a pilot effort aimed at understanding the underlying dimensions of social vulnerability, Cutter et al. (2003) created a list of population characteristics influencing social vulnerability. Using socioeconomic and housing data from the 1990 U.S. Census, they were able to explain more than 80 percent of the variation in vulnerability among all coastal counties with given population characteristics. This analysis provided much-needed empirical support for establishment of key indicators of vulnerability, or, those population characteristics that enhance or constrain the vulnerability of the human and built environment at the local level (Heinz Center 2002). When the factors are summed in a simple additive model, an overall composite score or social vulnerability index can be computed for each country. Nationally, the top five counties most socially vulnerable to coastal hazards are Cameron and Willacy, Texas; Manhattan, New York; San Francisco, California; and Charles City, Virginia, along the James River (Cutter et al. 2003). Cameron and Willacy counties in southern Texas are notable for poverty, racial mix, age, structure, and unemployment statistics. Manhattan and San Francisco have high values because of extensive development, and Charles City has high debt and employment reliance on infrastructure (transportation and utilities).

The SOVI is being used as part of the Coastal Services Center, Vulnerability Assessment Techniques and Applications workshops and is gaining some credence in the Caribbean. Some researchers investigating the human dimensions of global environmental change are using the SOVI in their work.

The SOVI provides a robust and consistent indicator for vulnerable populations and places and is useful in determining comparative levels of social vulnerability among places.

**Box 4.4   Community Development of an Information
Management Infrastructure for Southeast
Atlantic Ocean Coastal Observing System**

This program involves the development of a distributed, Internet-
based problem-solving community that is collectively focused on
developing the processes and tools to access, share, and integrate
heterogeneous data from multiple institutions and observing platforms. The community consists of about twenty-five data managers, information specialists, and computer-savvy technicians who
communicate largely through Internet-based e-mail threads and
forums.

As part of the emerging development of a national coastal integrated ocean observing system (IOOS), five institutions with nongovernmental organizations or nascent observing systems committed to
work together to link their systems in a way that would enable the
integration of data to produce new data products. This program—
the Southeast Atlantic Coastal Ocean Observing System (SEA-COOS)—
was established to enable characterization and predictive capabilities for a broader geographical region than was possible for each
individual partner on its own. The staff at each institution that
worked with data management was identified and institutions were
put into contact with each other via the Internet. This was quickly
followed by a two-day workshop in which they were tasked with
identifying their collective and individual capabilities, their near-
term needs for and obstacles to transferring information, and their

movement toward provision of real-time or near real-time in-situ and
remotely sensed data—has increased our ability to assess current conditions and facilitate rapid response to adverse changes. Numerous monitoring devices in coastal and ocean waters are now sending data back to a
central processing location via cell phone, submarine cave, or satellite
telemetry. Real-time data can then be processed, analyzed, and incorporated into models or computer visualizations, enabling "nowcasts" of current conditions. Such immediate understanding of current conditions was
simply not possible through the traditional delayed sampling and analysis
approaches. In many cases, models can also be used to provide predictions
of future conditions.

An example of an expansive infrastructure intended to provide
such real-time data on coastal and ocean conditions is the emerging integrated ocean observing system (IOOS), which has both global and coastal

Box 4.4 continued

potential solutions and approaches. The group assigned tasks among the participating institutions, identified priorities, and set timelines for accomplishing the next steps required for information transfer. This workshop set the stage for productive and very active communication via the Internet, in which a variety of problems were attacked and solved by collective action. Multiple subprojects were pursued and order was maintained by establishing independent discussion threads on separate topics. The participating group covers a four-state region. Additional links for solving problems or exchanging information have been made outside the region through connections made at workshops or through the Internet. Much of the Internet-based information exchange involves exchange among the entire group, while subgroups have been established to focus on specific observation infrastructure, data management, modeling, or outreach applications (see http://www.seacoos.org).

The principal lesson of this project to date is that the Internet is a powerful platform for exchange of information and for working through complex problems. Nonetheless, some face-to-face contact is important, so that participants can establish familiarity and trust with one another.

SEA-COOS Principal Investigator: Dr. Harvey Seim, Marine Sciences Department, University of North Carolina–Chapel Hill, Chapel Hill, NC 27599, e-mail harvey_seim@unc.edu. Data Management Working Group Chair: Dr. Madilyn Fletcher, Baruch Institute for Marine and Coastal Sciences, University of South Carolina, Columbia, SC 29208, e-mail fletcher@sc.edu.

components (www.ocean.us). Numerous observing systems have been established to address regional needs or scientific priorities, and these are now becoming coordinated to establish an integrated system in which information and information products can be merged and shared in an unprecedented manner (Box 4.4). The Ocean.US office was established to assist planning and implementation of this effort; it has outlined the observing system components, priorities, challenges, and applications (www. ocean.us). The IOOS approach and data management infrastructure will provide new access to coastal information, as well as new approaches and enhancements for obtaining and managing coastal and estuarine data.

In addition to local measurements provided by sensors on moorings, drifters, or submersibles, observations are also provided on global and regional scales by aerial and satellite imagery, such as the Advanced Very High Resolution Radiometer (AVHRR) data on sea surface temperature and Sea-viewing Wide Field-of-View Sensor (SeaWiFS) information on sea surface temperature and water quality (Jensen 1996). The potential of satellite imagery to improve understanding of coastal processes has grown with the advent of new hyperspectral or multispectral systems, such as Moderate Resolution Imaging Spectroradiometer (MODIS), as well as advances in the development of algorithms to analyze spectral data. Such data are exceptional in their ability to provide information on a global scale and often involve high volumes of data and computer capacity (Jensen 1996, 2000).

Communities of information users have responded to the new technologies and increasing volume of information by demanding and designing "user-friendly" systems and approaches. Easily navigated computer tools, such as Web browsers, can make information available to almost anyone. Translation tools that transform subject-specific data to alternative formats can facilitate use by the general user. Furthermore, commercially available software programs can facilitate information management and analysis by many diverse users and provide a common, accessible platform for a specific function, such as database management or geospatial analysis. Such software products enable integration of information and sharing of analysis products, usually after some initial orientation and training on the product, thus providing "interoperability" among distributed users. Increasingly, however, open-source software (freeware) solutions are being developed, which encourages the use of common inputs and outputs across diverse and distributed communities (see Open Source Initiative: http://www.opensource.org/). Open-source software development also taps into a creative and resourceful developer/user base that continually provides software enhancements, advances, and support, but generally requires some level of in-house technical expertise. The desirability of commercial versus open-source solutions is currently a subject of much discussion, and the costs and benefits of each, in the context of the organization's technical capabilities, should be considered when choosing appropriate approaches and investments in resources.

Users and user organizations are becoming more technologically savvy and willing to invest (at varying levels) in maintaining information technology expertise to optimize utilization of computer-accessed information. The private sector has generally recognized the value of technology training and included it as a strategic component for addressing corporate goals (U.S. General Accounting Office 2003).

Communication technologies are also increasingly used to facilitate interactions and information exchange (Schreiber 1998). Video conferencing helps to reinforce the face-to-face element of communication, which appears to be an essential contribution to efficient, sustained dialogue. It also is becoming increasingly accessible and affordable, particularly with teleconferencing based on high-bandwidth Internet connections rather than telephone lines. Many states, universities, and distributed organizations (e.g., companies with branch offices) regularly use video conferencing for meetings, instruction, and discussions. Systems can be inclusive, linking multiple local, national, or international sites with large screen monitors, or they can be as simple as a small camera connected to a desktop computer and accessed over the Internet. Community networks are becoming larger and more interconnected, thereby enhancing information sharing and training on community-relevant issues and information-sharing technologies (Box 4.5). While present-day technologies are providing creative and productive means for accessing the information needed to address specific problems, technologies of the future will provide greater promise for information sharing. They will also need user-friendly mechanisms for putting improved tools and information into the hands of the user community.

TECHNOLOGY'S INFLUENCE ON COASTAL MANAGEMENT PRACTICE

As new technologies have emerged, individual and organizational practices have also changed to realize the extraordinary potential provided by such revolutionary developments. The Internet, for example, has quickly become a key factor in transferring innovation in coastal management practices from one state or region to another. Indeed, a plethora of coastal information repositories are available via the World Wide Web at local, state, and regional scales. These include information on regulatory approaches, best management practices, GIS maps and interactive map services, workshops for coastal decision makers, and case studies.

In some regions, the emergence of coastal ocean observing systems (http://www.ocean.us) provides access to real-time and near real-time information that ultimately can be used to inform environmental decision making. Science-based information from these observing systems is typically accompanied by metadata, which detail information describing the data, such as where, when, how, and by whom the data were collected. Metadata

Box 4.5   Connecticut's NEMO Project Used as Model Nationwide

Nonpoint Education for Municipal Officials (NEMO) is an educational program that helps community decision makers protect their natural resources, while charting the future course of their towns. Created in 1991 by the University of Connecticut Cooperative Extension Service (UConn/CES), in partnership with the Department of Natural Resources Management and Engineering and the Connecticut Sea Grant Program, NEMO receives funding from a number of federal and state agencies. Major funding is provided by the U.S. Department of Agriculture's Cooperative Research, Education, and Extension Service Water Quality Program, the University of Connecticut, the Connecticut Department of Environmental Protection (CT DEP), the National Oceanic and Atmospheric Administration, and the U.S. Environmental Protection Agency.

The program was stimulated by a need to better explain the links between land use, water quality, and community character to land-use decision makers. It uses advanced technologies—geographic information systems, remote sensing, and the Internet—to create effective education programs. NEMO presentations, publications, and Web-based services integrate information around the theme of planning for sustainable use of natural resources. The program offers follow-up presentations and materials to help communities move forward on the two major aspects of natural resource planning, namely, planning for areas to be preserved and planning for areas already developed or developing.

In 1997 CT DEP awarded Section 319 grant funds to NEMO to expand its program of technical assistance for local officials. During the first year, NEMO delivered its basic presentation through a series of ten regional workshops. More than 120 of the state's 169 municipalities were represented at the workshops, and many participants contacted NEMO to schedule follow-up meetings on specific issues or concerns. Each municipality also received a map set (watersheds and land cover) to help educate local officials and facilitate non–point source management at the local level. In 1998 and 1999 NEMO conducted regional workshops to teach local officials how to manage non–point source pollution through their land-use planning and regulatory authorities. Since 2000, the program has continued to conduct regional workshops for new land-use commissioners, but has also moved to a more intensive approach, selecting on a competitive basis five communities each year to enter the Municipal Program. Each of the communities is charged with developing specific goals, creating a NEMO committee made up of representatives from all the land-use boards and commissions and other interested parties, and designating a chief NEMO contact to facilitate progress.

After almost nine years of experience with the NEMO program, Connecticut municipalities are giving greater consideration than ever before to water quality in their land-use planning and regula-

Box 4.5 continued

tory programs. Two examples provide concrete evidence of NEMO's effectiveness.

■ As a result of NEMO's Eightmile River Watershed Project, the towns of Lyme, East Haddam, and Salem signed the Eightmile River Watershed Conservation Compact, which commits the towns to work together to protect natural resources from new development. Since the signing, the three towns, local land trusts, and The Nature Conservancy have protected more than 1,800 acres of open space in the watershed. In addition, UConn/CES foresters have worked with landowners to develop forest stewardship plans on almost 500 acres and provided information that is being used to manage another 2,500 acres of forestland. The program was also instrumental in helping to build a fish ladder to restore access to upstream habitat for alewives and blueback herring for the first time since the early 1700s.

■ As one of NEMO's original pilot projects, the suburban coastal municipality of Old Saybrook has a long-term relationship with the project that has resulted in a progression of positive impacts. The zoning commission reduced the number of required parking spaces in several site plans, where it could be demonstrated that fewer cars were likely, to reduce the amount of surface impervious to runoff during rainstorms. Associated landscaping regulations were revised to require the breaking up of "seas of asphalt" through the use of landscaped islands and buffers. The Conservation Commission revised the town's conservation plan to include a recommendation on controlling non–point source pollution and recently completed a natural resources inventory for the town. The Board of Selectmen prepared a policy statement that includes alternative design and construction standards and vegetative storm water management practices that were incorporated directly from NEMO program design principles and are in keeping with Phase II storm water permit requirements.

Based on the success of the first several years of this partnership, CT DEP anticipates continuing its support for NEMO and now considers NEMO an integral part of the state's Non–point Source Management Program. NEMO is continuing its Municipal Program, as well as impervious surface research.

The NEMO Program at the University of Connecticut is the coordinating center for the National NEMO Network, a growing network of projects around the country adapted from the Connecticut project. As a result of NEMO's success in Connecticut, thirty-four other states have established or are planning to establish technical assistance programs based on the NEMO model. For more information about the NEMO Program, visit http://nemo.uconn.edu.

also include information that helps to assess data quality, such as cautionary notes about suspect data, data gaps, or data collection anomalies. Equally, the absence of particular kinds of metadata also indicates data quality.

Such evaluation information is generally lacking for the management-oriented repositories available to the coastal management community. Reference to credible sources, such as peer-reviewed publications or reports from well-established research centers, instills some confidence in the information. Nonetheless, without background information or metadata about Web-based coastal management innovations, users must establish the credibility of the information elsewhere, usually from a trusted source within an established communication network.

Thus, even while technology is transforming the way coastal management information is gathered and displayed, personal interaction remains absolutely essential for the transfer of innovations. Nothing replaces human involvement in the development of sound resource management decisions. Coastal management requires the management of people as well as habitat. With much coastal decision making occurring at the local level, strategies to engage specific stakeholder groups are a key to successful innovation. However, this type of information is more readily transferred personally rather than electronically.

MOVING FROM DATA TO USEFUL INFORMATION

The extraordinary increase in available data has brought with it a major challenge—how to process the data to provide sound and useful information. The first issue to consider is how to treat new data and assimilate it into the context of present knowledge to provide a new understanding of the problem, situation, or concept. Only this kind of new understanding, based on sound information, can influence human behavior to modify or institute management practices.

A second issue to address is the quality of data or information, including its relevancy, accuracy, and resolution. With Web-derived material, this can be a difficult task. Because of the need to assess data quality, it is imperative that metadata are included with data and are expressed in commonly understandable terms. Adherence to those standards specified by the Federal Geographic Data Committee (http://www.fgdc.gov/) is an essential first step, but further developments and enhancements are needed to develop metadata standards for the extraordinary variety of information relating to coastal zone management.

Nevertheless, many types of information will never be discretely defined data with accompanying metadata, and for these, personal knowledge of the credibility and reliability of the source plays a primary role in judging quality. In coastal management, like other science-based but human-oriented decision-making processes, the onus is on the user of data to assess their reliability and ultimate usefulness.

FUTURE TECHNOLOGIES AND COASTAL MANAGEMENT

The development of new and enhanced technologies is likely to continue at a rapid pace, as the Internet is integrated with wireless communications, new telephone technologies, miniaturization and nanotechnologies, artificial intelligence, software agents (bots), biotechnology, and ever-increased raw computing power. Much of this technological development will be motivated, tested, and implemented within the private sector, which continues to play a critical role in the innovation and adaptation of technology to creative uses.

A number of advances are already under way with the Internet. For example, a second-generation Internet infrastructure—Internet2 (http://www.internet2.edu)—is being established by the university community, in association with industry and the federal government, outside of the increasingly cluttered current Internet. Internet2 operates at a higher bandwidth than the Internet and will connect specific university and government users. Even if coastal managers do not see a direct benefit from Internet2, the new system will provide a fertile field for problem solving and development and will help to advance Internet technologies and applications beyond their current framework. There will also be an increased decentralization of Internet communications. Wireless technologies are releasing users from their desktops and providing continuous access to the Internet for travelers and people working in the field. There will also be a decrease in the single server–client relationship, with more distribution of databases and sharing of computer and information resources among peers and computing networks (Foster 2000).

New ground is also being broken in the software used for Internet applications. The increasing use of open-source software, already noted, allows a bottom-up approach to problem solving throughout a broad community. Another approach may be the increasing use of software to write new software, an approach termed genetic programming, in which the human programmer guides the process toward the desired result (see http://www.genetic-programming.org/).

Transferring information is a technology challenge, but delivering that information in a manner that has maximum impact is a communication challenge, one that requires attention to the ways in which humans perceive and assimilate information. Artificial intelligence, technology that provides seemingly human-like cognitive abilities, may have the potential to relate with users more effectively than the current generation of desktop software. For example, an "intelligent agent" can respond to specific questions in an interactive manner or may even be tasked with routine jobs, thus providing more time for more creative activity by the human operator. The presentation of information in a visual form (with GIS software, for example) is often effective for conveying complex, multi-component, or time-dependent information quickly and dramatically. Such visualizations are assimilated via visual pattern recognition and spatial memory, rather than standard written or numeric cognition, and hence may be more effectively perceived and learned.

Although this report cannot analyze each of these emerging technologies, we can try to predict the future use of information and communication technologies in coastal management through an extrapolation of recent trends in innovation. One clear implication for the practice of coastal management is that the technologies available to professional managers will dramatically change in the future. With costs falling and accessibility increasing, today's specialist management tools are likely to become more widely available to nonprofessional stakeholders, including a variety of coastal users. Another clear trend is an increasing awareness of the "emergent properties" of innovation, in that new and perhaps unexpected applications arise from current patterns of development. For example, new technologies and services might emerge by combining streams of innovation, or they could emerge from the nexus between the technology and its societal use. Because these changes are unforeseen, they will require management when and where they occur. As a result, one of the critical issues for coastal management is to develop governance arrangements that permit flexibility in adapting to technologies (and their unintended consequences) as they are developed.

INCORPORATING TECHNOLOGIES
INTO ORGANIZATIONAL PRACTICES

If innovations in information and communication technologies are likely to continue at a rapid rate, how can these best be incorporated into practice to improve coastal management? Although the learning networks available to coastal managers are developing an increasing knowledge base for accessing and sharing relevant data and information, the technological knowledge base of the community has not, and likely cannot, keep pace with that of the technology developers. New technology can be intimidating, even for the savvy. Moreover, the time it takes to become aware of and proficient with a new technology can be prohibitive. And, of course, proficiency by one member of a learning community has limited value—genuine value only comes when a large part of that same learning community is also proficient and engaged with a particular technology tool.

Further delays result from the persistent gap in communication between the technology developers and the various users who could benefit from ready access to computer-based information. The two communities have different expertise, different language sets, and even different cultures. These differences have led to a perceived dichotomy in how to introduce new technologies into the user organization's repertoire. On one hand, some organizations have invested in in-house information specialists who can handle information-technology activities. On the other, some developers strive to design information systems so they are user friendly, simple to install, and easy to understand and navigate. In practice, an integration of these two approaches will bring most success in managing information.

Information specialists have a fundamental role in an organization and should work cooperatively with, and be accessible to, all organizational staff members. The incorporation of technical specialists into an organization that has never had them may be difficult to achieve, because new positions require additional resources, which are often in short supply. Nonetheless, organizational leaders should recognize the value that technical specialists within the core framework can add by making less technically oriented staff more comfortable with technologies. A viable and often cost-effective alternative to in-house expertise is to use outside consultants for securing technical proficiency and training. In either approach, leaders will find that technologies become integrated into the work community, ensuring optimum utility and impact. With familiarization, staff become less intimidated by technologies and more proficient and capable of making decisions about which tools to use and for what purposes.

Specific training programs to inform workers about available technologies are extremely effective, as they provide a focused, often personalized environment in which to learn new concepts, techniques, and applications. The personal interaction achieved through conferences and workshops are particularly effective for transferring new technical information and techniques. Remote training platforms, such as Web portals and increasingly affordable video conferencing, are also useful because they are easily accessed and eliminate the time and cost required for travel to offsite training (Schreiber 1998). Regional onsite training programs are also very productive, as they provide access to a high level of expertise in a personally interactive environment, without the expense of hiring in-house expertise.

Even with the best intentions, organizations attempting to incorporate new, sophisticated technologies into their routine practices and applications will sometimes face extreme difficulty. A commitment must be made to the transition, and appropriate effort and resources must be allocated. Technology rarely provides a "quick fix," and technology solutions are as complicated as the problems they address. For example, many organizations are attempting to substitute program management software for well-entrenched paper-based processes. The National Science Foundations' FastLane, (https://www.fastlane.nsf.gov/fastlane.jsp) (Box 4.6) which is now the basis for essentially all grant proposal submission, review, and communication processes, works effectively and deals with a high volume of complex information. However, the system required years to implement and initially contained many "kinks" that had to be identified in practice and subsequently addressed.

Similarly, the Office of Ocean and Coastal Resources Management (OCRM) has begun to develop the Coastal and Marine Management Program (CAMMP) to manage its grant application process. Its full implementation has been delayed by technical issues, which are being addressed in a more advanced product. Like FastLane, the CAMMP system has taken years to refine and implement (Box 4.7).

Box 4.6   The National Science Foundation's FastLane

To streamline business interactions with the research and education communities by using advanced information technology, the National Science Foundation (NSF) established FastLane, an electronic information system to manage proposals and project information. FastLane was developed in 1994, and by 1998 it was implemented by the research community (https://www.fastlane.nsf.gov/a0/about/fastlane_ history.htm).

With the advent of interdisciplinary studies involving many investigators from different institutions, proposal writing and peer review became much more complicated and time-consuming. Principal investigators were required to submit ten to twelve copies of proposals, thereby deluging NSF with hard copies of grant applications. Proposals then had to be mailed out to individual reviewers with evaluations returned in the same manner. A great deal of time and funds were being expended to manage the peer review process. Users seeking information on funded projects had difficulty identifying relevant investigators or had to speak directly with program managers.

With FastLane, principal investigators can prepare and submit proposals online, and project managers can use the system to track project progress. In addition, reviewers can evaluate proposals electronically. As a result, doing business with NSF has become "simpler, faster, more accurate, and less expensive" (https://www.fastlane. nsf.gov/fastlane.jsp). FastLane merits consideration as a well-designed, common interface for the coastal zone community to interact with the federal government.

NSF initially selected a few programs to pilot FastLane. This enabled the system to be debugged and improved before it was implemented fully. Now, all investigators submitting proposals or filing project reports with NSF are required to use FastLane.

When first piloted, NSF was deluged with calls for assistance just prior to proposal deadlines. In addition, the computer system was unable to handle the quantity of proposals submitted at the deadline, thus requiring the deadline to be extended. An investment in computer support and personal assistance and the establishment of a help-line were critical to the successful launch of this information management system.

**Box 4.7   CAMMP: Results Accounting for State Coastal
Management Programs**

In 1996, the Office of Ocean and Coastal Resources Management
(OCRM) began to develop the Coastal and Marine Management Program (CAMMP) information system to manage its grant applications.
OCRM administers cooperative agreements with state coastal management programs and estuarine research reserves. Each coastal
management program and estuarine research reserve is different,
and before CAMMP the format and content of each application varied
widely. CAMMP standardized grant applications by creating a structure that all programs must use to create their applications. (Grant
applications are composed of tasks which each have a budget and at
least one outcome.) All information that is entered into a grant
application goes into a database, where it can later be queried,
retrieved, and analyzed. The system also improved applications by
condensing task and outcome descriptions. Even after the first version of the program was retired because of technical problems,
states have continued to use the CAMMP format.

   CAMMP was created to allow OCRM and its programs better
access to programmatic data in the applications, to eliminate errors
that slow grant application approval, and to facilitate comparisons
across programs by standardizing formats. The CAMMP system is also
enabling OCRM to better meet the federal government's automated
grant requirements.

   Early versions of CAMMP have allowed OCRM to capture information about the state programs more efficiently and effectively.
The latest version of CAMMP, which should be available in early
2004, is expected to build on this success. Future refinements will
make the system even more useful. A planned query function will
allow state program managers and OCRM staff to search for particular kinds of projects by state or by fiscal year. For example, an OCRM
staffer writing the Biennial Report to Congress might be able to
quickly locate all the public access projects planned for the previous
year, simplifying reporting requirements. A state coastal program
manager interested in designing a habitat restoration project might
be able to search for similar projects that had been conducted in
neighboring states.

   A second planned improvement is to link information in CAMMP
to performance indicators being developed for the national performance measurement system. For example, if a wetland restoration
indicator is selected for the national system, OCRM staff would be
able to query for the number of acres of wetlands restored in a given
year. In the longer term, if funding permits further development,
states will be able to post program changes and reports to the

Box 4.7 continued

CAMMP system. Automating all reporting and applications would streamline the process for both applicants and state liaisons. Program analysis would also be easier because OCRM staff would be able to review not only proposed projects, but also what actually happened.

CAMMP will be voluntary for states in fiscal year 2004. Once the program has been tested and refined, OCRM may require all states to submit their grant applications to CAMMP. The National Oceanic and Atmospheric Administration (NOAA) may also adopt the CAMMP model for its other noncompetitive, nondiscretionary cooperative agreement applications.

What has OCRM learned from this process? Designing and successfully implementing an information system is a complicated endeavor, which requires partners, communication, and the proper design to succeed. The first version of CAMMP was unsuccessful because it was developed using customized software that could not be adapted over time without programming skills that OCRM lacked internally. The second version failed because of unrealistic scheduling and inadequate testing of the system before it went live.

This latest version of CAMMP has tried to avoid all these problems by bringing in partners from other parts of NOAA to provide the technical expertise to support and manage the system. In addition, the system has been designed, developed, and tested over the course of a year; management has participated in weekly status meetings; and the system will not be allowed to go "live" until it has passed all the planned tests.

Because the current version of CAMMP is still under development, the program's greatest innovations are perhaps yet to come. Yet, even initial versions of the program have demonstrated that there are enough commonalities among the diverse coastal programs to standardize grant applications and reporting. Collecting program data in a database will allow much more efficient and accurate analysis and reporting on program activities and successes. For more information, contact Dwight Reynolds at 301-713-3155 x154 or dwight.reynolds@noaa.gov.

CENTRALIZED STANDARDS, DISTRIBUTED KNOWLEDGE

Coastal managers in our survey have already recognized some basic needs that should be met by any new technology: they want clear routes of access and clear standards of comparison for all the available data and good ideas. To ensure this access and standardization, a centralized—

probably federal—system is required to establish some common defini-
tions and formats that will be used by all of the relevant organizations
providing information. A central system can establish consistency among
ways in which different data are measured, recorded, and described.

The National Spatial Data Infrastructure (NSDI) has been established to provide such consistency for sharing geospatial data, in order to
increase accessibility to information, reduce duplication, optimize costs,
and facilitate partnerships (http://www.fgdc.gov/nsdi/nsdi.html/). The
NSDI is being developed by the Federal Geographic Data Committee
(http://www.fgdc.gov/) and includes the Office of Management and Budget e-government initiative Geospatial One-Stop (http://www.fgdc.gov/
geo-one-stop/), which is tasked with providing a geographic component
for Internet-based e-government activities across multiple government
sectors. Additional efforts to establish and disseminate standards include
the U.S. Geological Survey's National Map (http://nationalmap.usgs.gov/),
an online interactive map service, and the user-driven Open GIS Consortium, a nonprofit international trade association that is working to
develop open software applications for geospatial and advanced technology interoperability (http://www.opengis.org/). At the state level, the
National State Geographic Information Council (NSGIC) is an organization of state GIS coordinators and senior state GIS managers, as well as
other representatives from government, academia, the private sector, and
professional organizations, which promotes effective and efficient government through geographic information technologies (http://www.nsgic.
org/). These diverse initiatives all demonstrate the growing recognition of
the need and potential for common standards and formats for information management.

Many coastal managers have expressed a wish for the reestablish-
ment of the regional Coastal Information Centers of the early 1980s,
which would provide a central repository for archiving and providing
information. Thus, managers would know where to go for reliable and
comprehensive information, without having to search for distributed
and heterogeneous data. However, such government-organized databases
can never wholly fulfill the need for information sharing. Data may be
best accommodated by such systems, but much of what coastal managers
and professionals are looking for are good ideas and programs they could
emulate or modify to address local issues. For that purpose, a less data-
oriented system is needed, one that accommodates ideas, details, contacts,
and discussion. Thus, the ideal information-sharing technology needs not

only a site where information is collected, but also some mode of nongovernmental organization communication and contact.

CONCLUSIONS AND RECOMMENDATIONS

The new communication, analytical, and observing technologies have the potential to revolutionize how we access, use, and implement information for improved resource management. However, the coastal management community lags behind many constituencies in taking advantage of this new information trove. Many community members have not had the opportunity to embrace the new technologies and incorporate them into their operations and decision-making process. Progress is needed both in familiarizing the community with the tools at hand and in developing a new generation of tools that are more user-friendly and intuitive. Moreover, it is important to realize that successful adoption of new technology can be a difficult transition, requiring an acceptance of change and a commitment of effort and resources. Regardless of the many opportunities and efficiencies that can be provided by new technologies, personal interactions and communication are still essential. Personal exchange and networking within communities build the mutual trust required for true information sharing. Personal interaction is also needed to track the implementation of new technologies and assess their effectiveness in solving real-world problems.

Several recommendations are suggested by this analysis of technology and coastal management practice. First, NOAA should collaborate with other agencies to institutionalize a learning process about interactions between technology and coast management—a periodic national workshop or conference emphasizing different technologies would be an effective way to achieve this. As noted in Chapter 3, NOAA, the USEPA, and other federal agencies should expand the cross-training of personnel to broaden mutual understanding of coastal management problems, practices, and use of technology. Finally, the coastal management community should utilize existing and always improving technology to distribute workshop and conference sessions to broader regional audiences, for example, through Webcasting.

APPENDIXES

Appendix A

THE SURVEY AND
INTERVIEW PROCESS

THE HEINZ CENTER for Science, Economics and the Environment, a private, nonprofit public policy research center in Washington, D.C., has been contracted by the NOAA Office of Ocean and Coastal Resources Management (OCRM) to explore how good ideas and practices for coastal problem solving are shared with others who have similar needs, and how information sharing might be improved. The Heinz Center has brought together a dozen experts in coastal management from government, academia, business, and nongovernmental organizations to undertake the study.

The impetus for the study comes from a number of sources. OCRM regularly receives queries from state coastal managers wanting to find out how others have addressed the problems they are facing, but has no systematic means of providing this information in a comprehensive manner. In the 1999 NOAA Coastal Services Center's "customer survey," 80 percent of respondents said they needed better "access to information about and how other offices have addressed similar issues and management options." Clearly, coastal managers are saying they need to become a more effective and efficient learning community.

Our study addresses this need, with specific objectives to

1. Define the problem more clearly by documenting how we *do* share coastal problem-solving ideas and practices now in government, academic, business, and not-for-profit sectors.
2. Identify the strengths, limitations, and outlook for present information-sharing methods and efforts.

3. Identify ways to improve information sharing and learning, drawing on the experiences of those within and outside the coastal management community.

THE STUDY PROCESS

To begin our study, we want to learn from those involved in coastal management—planners, regulators, developers, and advocates in the public, academic, private, and not-for-profit sectors. Our first information-gathering effort will involve semi-structured interviews of coastal managers in government, academia, and the private and nongovernmental sectors. Following our interviews, we plan to conduct a small workshop to explore in more depth what we learn in the interviews. We will invite a number of coastal managers to the workshop, as well as others who can help us learn more about becoming a more effective learning community. The panel will then deliberate and prepare its report to NOAA.

SURVEY/INTERVIEW

The interview questions below are designed to facilitate our learning. If we haven't asked the right question, tell us so. We need to know what you think about, How can coastal managers and decision makers learn and adapt the lessons and experiences of others to achieve better outcomes?

Confidentiality. Although we may generalize results of this survey, individual interviews are confidential. We will ask your permission if we wish to cite specific examples you provide.

INTERVIEWEE   INTERVIEWER
Name: Name:
Address:
Phone:
E-mail:

Interview Questions
1. Looking back over the last decade or so, what are the most important changes or improvements in coastal programs, planning and development processes, regulatory or acquisition programs, etc., that have affected how you do business? [*Examples include (1) a new way to track*

and report performance, (2) new way of assessing wetland functions and development tradeoffs, (3) improving public access to beaches or waters, (4) cleaning up and redeveloping brownfields or deteriorated waterfronts, (5) involving the public and interest groups in decision making, (6) public-private collaborations, (7) addressing sea level rise within bays and estuaries or along the ocean coasts, and so on.]

Possible Follow-up Questions:
A. What were the motivations for making these changes or improvements? [For example, a directive from higher up, a response to other pressures, a clear resource or development-related problem, etc.]
B. For state CZM staff—in the overall scheme of problems you deal with, how important are national initiatives as drivers for changes or innovations, for example, the 6217 non-point pollution program or the CZM Section 309 CZM improvement process over the last decade? To what extent have 309 financial resource availability or the skills of staff and leaders played roles in the outcomes of initiatives you've undertaken?
C. Were the changes or improvements you made based on conventional practices, or would you characterize them more in the "innovative" category? [We define innovative as simply an idea or practice that you perceive as *new*.]

2. Where did you turn for information, ideas, strategies, or tools to help design and implement changes and improvements in coastal management? What are the most important and credible *channels* of communication?

Possible Follow-up Questions:
A. Is the search process and information availability issue-dependent?
B. In designing your program improvements, were there good examples elsewhere from which you could draw lessons?
C. Did you have to reinvent or tailor existing approaches in significant ways? How did you go about that?
D. How important is networking for getting problem-solving ideas or approaches?
E. What individuals, organizations, or media do you (and your staff or colleagues) most rely upon?

   F. Do you ever bring someone in from another state, agency, or program to give you the details first hand and help you think through the adaptation process?
   G. Were there situations where you could find nothing useful elsewhere and had to work from scratch, inventing your own approach and solution?

3. How do the really good ideas and practices in coastal management spread or diffuse? Can you give an example?

   Possible Follow-up Questions:
   A. How has technology fostered innovations in problem solving and in getting innovative solutions to problems transferred to others who could use them?
   B. What kinds of information transfer media work best for you—for example, conference presentations, in-depth workshops, project reports documenting processes and results, case studies, demonstration projects, journal articles, CDs, published reports, and so on?
   C. How do you share your good ideas and practices for coastal management problem solving with others? Can you give an example?

4. How would you like to get your information in the future, given the constraints you face and the resources available—human, financial, and technological? What should be given more emphasis by whom and why?

5. Finally, are there other people you think we should interview to learn specifics about a particular information sharing example or coastal management innovation?

Thanks very much for your time and ideas. We will keep you abreast of the study process and put you on the mailing list for our final report. In the meantime, if you have anything you would like to follow up on, don't hesitate to call or e-mail. You can contact me or the study coordinator, Sheila David at The Heinz Center (202-737-6307 sdavid@heinzctr.org).

Appendix B

Survey Participants

Ray Allen, Coastal Bend Bays and Estuaries Program, Texas
Wendy Allen, North Inlet-Winyah Bay NERR, South Carolina
Robert Bailey, State of Oregon Coastal Management
Brad Barr, NOAA National Marine Sanctuaries, Woods Hole, Massachusetts
Lillian Borrone, President's Ocean Policy Commission, Washington, D.C.
Jeb Boyt, Coastal Management Program, Texas General Land Office
Doug Canning, Coastal Management Specialist, Washington State Department
  of Ecology
David Carter, State of Delaware Coastal Management
Chris Chung, Office of State Planning, Hawaii Coastal Management
Elizabeth Corbin, Hawaii Department of Business, Economic Development,
  and Tourism
Tim Dillingham, American Littoral Society, New Jersey
Helen Drummond, Galveston Bay National Estuary Program, Texas
Richard Eckenrod, Tampa Bay National Estuary Program, Florida
Robert Goodwin, Washington Sea Grant, Washington
Mike Graybill, South Slough NERR, Oregon
Bryon Griffith, USEPA, Gulf of Mexico Program Office, Mississippi
Mike Guilbeaux, Community Conservation Network, Hawaii
Debbie Heaton, Sierra Club, Delaware
Ginger Hinchcliff, NOAA Coastal Services Center, South Carolina
Phillip Hinesley, State of Alabama Coastal Management
Fred Holland, Hollings Marine Lab, South Carolina
David Keeley, State of Maine Coastal Management
Geraldene Knatz, Port of Long Beach, California
Kathleen Leyden, State of Maine Coastal Management
Gary Lytton, Rookery Bay NERR, Florida
Fred McManus, USEPA, Florida Keys Water Quality Program, Florida

Sam Messina, State of New York Coastal Management
Jeffrey Pantazes, Public Service Electric and Gas, New Jersey
Dwayne Porter, Baruch Institute, University of South Carolina
Peter Rappa, Sea Grant, University of Hawaii
Rebecca Roth, California Coastal Commission
Paul Sandifer, Hollings Marine Lab, South Carolina
Terry Steven, Padilla Bay NERR, Washington
Bob Tudor, Delaware River and Bay Commission, New Jersey
Tom Wakeman, Port Authority of New York and New Jersey

Appendix C

WORKSHOP PARTICIPANTS

The Heinz Center workshop was held in Woods Hole, Massachusetts, June 25–27, 2003.

David Bancroft, Alliance for the Chesapeake Bay, Maryland
Betsy Blair, Hudson River NERR, New York
Doug Brown, NOAA, Office of Ocean and Coastal Resource Management
William Burgess, Maryland Department of Natural Resources
Ralph Cantral, NOAA, Office of Ocean and Coastal Resource
Jeanne Christie, Association of State Wetland Managers, New York
Charles Colgan, University of Southern Maine
Sarah W. Cooksey, Delaware Coastal Programs
Tracey Crago, Woods Hole Oceanographic Institute Sea Grant, Massachusetts
Sheila D. David, The Heinz Center, Washington, D.C.
Richard Delaney, Urban Harbors Institute, Massachusetts
Michael Deluca, Rutgers University, New Jersey
William DuPaul, Virginia Sea Grant
Dick Eckenrod, Tampa Bay NEP, Florida
Jon Fisher, Texas Chemical Council
Madilyn Fletcher, Baruch Institute, University of South Carolina
Christine Gault, Waquoit Bay NERR, Massachusetts
Jim Good, Oregon State University
Judy Goss, The Heinz Center, Washington, D.C.
Dawn Hamilton, The Coast Alliance, Washington, D.C.
Olwen Huxley, U.S. House Science Committee, ETS Subcommittee
Robert Kay, Kay Consulting, Mosman Park, Australia
Jack Kindinger, USGS, Center for Coastal and Watershed Studies
James Langdon, Wisconsin Coastal Programs

Kathleen Leyden, Maine State Coastal Programs
Kem Lowry, Jr., University of Hawaii
Tony MacDonald, Coastal States Organization, Washington, D.C.
Andrew Manus, Ducks Unlimited, Maryland
Kalle Matso, CICEET, New Hampshire
Donna McCaskill, NOAA, Coastal Services Center
Arleen O'Donnell, Massachusetts Department of Environmental Resources
Steve Olsen, Coastal Resources Center, University of Rhode Island
Rebecca Roth, California Coastal Commission
Miki Schmidt, NOAA, Coastal Services Center
Suzanne Schwartz, USEPA, Ocean and Coastal Protection Division
Nicholas Shufro, United States–Asia Environmental Partnership Program, Washington, D.C.
Susan Snow-Cotter, Massachusetts Coastal Programs
Caroline Stem, Foundations for Success, New York
Maya K. van Rossum, Delaware Riverkeeper Network
Robert Wayland, Consultant, Virginia

Appendix D

ABOUT THE CONTRIBUTORS
AND THE PROJECT STAFF

THE CONTRIBUTORS

JAMES GOOD, *Chair*, is professor and director of the graduate program in Marine Resource Management in the College of Oceanic and Atmospheric Sciences, Oregon State University. He has served as Oregon Sea Grant's Coastal Resources Specialist (1980–2003), conducting applied research and outreach programs for Oregon and the Pacific Northwest. Much of his teaching, research, and outreach work has focused on state and community-based planning and management for estuaries, urban waterfronts, beaches, and marine environments. He led the marine and estuarine ecosystem assessments for Oregon's State of the Environment Report 2000 and was principal investigator for estuaries and coastal wetlands in the National CZM Effectiveness Study. Prior to coming to Oregon State University in 1980, he was executive director for the Columbia River Estuary Study Taskforce and, before that, a U.S. naval officer. He received his B.A. in chemistry from Susquehanna University, M.S. in marine resource management, and Ph.D. in geography from Oregon State University.

RALPH CANTRAL is chief of the National Policy and Evaluation Division, Office of Ocean and Coastal Resource Management, National Oceanic and Atmospheric Administration National Ocean Service. He served as executive director of the Florida Coastal Management Program from 1992 through 2001. In 1999–2000, he also served as acting executive director of the Florida Communities Trust, a $66 million-per-year land acquisition program for local governments. Prior to moving to Florida, he served in a number of coastal management, planning, and community development positions with the State of North Carolina. He has also worked for local government and regional agencies. He serves on a number of state and national boards and committees, and is published in the fields of coastal management, planning, and dispute resolution. He holds B.A. and M.A. degrees in geography.

CHARLES COLGAN is professor of public policy and management in the Edmund S. Muskie School of Public Service at the University of Southern Maine. He is chair of the Muskie School's graduate program in community planning and development and is associate director of the USM Center for Business and Economic Research. His regular economic analysis activities include serving as the Maine model manager for the New England Economic Project and chair of the State of Maine Consensus Economic Forecasting Commission. His long-term economic forecasts are used by the Maine Department of Transportation and the economic development districts of Maine. Prior to joining the University of Southern Maine, he served with the Maine State Planning Office in the administrations of Governors James Longley, Joseph Brennan, and John McKernan. His state government positions included state economist, director of natural resource and economic policy, and special assistant to the governor for international trade policy. He also served as director of research for the Finance Authority of Maine. He received his B.A. from Colby College in 1971, did graduate studies in international relations at the University of Pennsylvania, and received his Ph.D. in economic history from the University of Maine in 1992.

SARAH W. COOKSEY has been involved in environmental protection for the past fifteen years. She has been head of the Delaware Coastal Programs for the past ten years, responsible for both the Delaware Coastal Management Program and the Delaware National Estuarine Research Reserve. She is past chair of the Coastal States Organization and currently a member of its executive committee. She serves as the coastal management representative to the Cooperative Institute for Coastal and Estuarine Environmental Technology Advisory Board. She was employed for a number of years by the U.S. Environmental Protection Agency in Washington, D.C., where she worked with state governments on water pollution control.

MICHAEL DELUCA received his B.S. in biology from Fairleigh Dickinson University in 1978, and M.S. in marine science from the Virginia Institute of Marine Science College of William and Mary in 1984. He currently serves as the senior associate director of the Institute of Marine and Coastal Sciences at Rutgers University. He is chair of the Rutgers Dive Safety Control Board and the director of the Mid-Atlantic Bight National Undersea Research Program. He is also the manager of the Mullica River–Great Bay National Estuarine Research Reserve. His research interests include environmental policy related to resource management (especially coastal, estuarine, and fishery management), science education and the promotion of environmental awareness among students and educators, watershed approaches to coastal ecosystem management, dredged material management and sediment decontamination, and undersea technology.

JON FISHER is senior vice president for Texas Chemical Council (TCC). He began at TCC in 1981 as director of research, and has held the titles of vice president–research and senior vice president–research. He served from 1994 until 2002 as vice president–government affairs for the Texas Ag Industries Association (TAIA), having served as vice president of one of its predecessor associations. In 2001, TAIA presented him with its Harry P. Whitworth Political Involvement Award. In 1989, he received an Outstanding Service Award from the Texas Agricultural Aviation Association (where he worked for eight years in the 1980s). At TCC, he has had a hand in drafting or working on every significant piece of environmental and hazard communication legislation in Texas since 1981. He is also active in lobbying at the regulatory level. He was the leader of the Chemical Industry Focus Group, which helped develop Texas' current Coastal Management Plan, and represents the chemical industry on the Gulf of Mexico Coalition Business Council.

MADILYN FLETCHER is currently the director of the Belle W. Baruch Institute for Marine Biology and Coastal Research at the University of South Carolina. The Baruch Institute develops and conducts programs in multidisciplinary research on coastal, estuarine, and marine systems (http://www.baruch.sc.edu). In her role as Baruch Institute director, she has a strong interest in regional partnering and the development of initiatives that coordinate strong science with real-world applications and needs. She is principal investigator for the Carolinas Coastal Ocean Observing System (Caro-COOPS), a new initiative with partners North Carolina State University and the University of North Carolina–Wilmington. Caro-COOPS is currently being implemented and will comprise a mooring array off of the Carolinas' coast, which is designed to integrate real-time monitoring of hydrologic and meteorological conditions with state-of-the-art computer models to characterize and predict complex coupled air-land-sea processes. The University of South Carolina portion of this collaborative project is focused on information management and integration. She is also principal investigator for Cast-Net, a multi-institutional program focused on development of tools to facilitate documentation, integration, and sharing of data from laboratories in the Southern Association of Marine Laboratories (SAML).

KEM LOWRY, JR., is professor and chair of the Department of Urban and Regional Planning, University of Hawaii. He received his Ph.D. from the Department of Political Science, University of Hawaii in 1976. He has been a visiting scholar at the Institute for International Relations and Development in Asia, Sophia University, Tokyo; visiting faculty at the Department of City and Regional

Planning, University of North Carolina; and a fellow at the Marine Policy Program, Woods Hole Oceanographic Institution. He has published articles on planning and environmental management, and coastal management and evaluation in numerous journals. He is also the co-author of *Choosing Change: A Self Assessment Manual for Non-Profits.*

ANDREW MANUS is currently director of conservation programs for the mid-Atlantic office of Ducks Unlimited, Inc. Prior to joining Ducks Unlimited, he served as director of the Delaware Division of Fish and Wildlife. During his tenure from 1993 to 2001, he represented the state of Delaware on the Atlantic States Marine Fisheries Commission, the Atlantic Flyway Council, the North American Wetlands Conservation Council, and chaired the International Association of Fish and Wildlife Agencies' Wildlife Diversity Committee. His state service has also included time as deputy director of the Divisions of Soil and Water and Water Resources within the Delaware Department of Natural Resources and Environmental Control. In that position he led the development and design of an "Environmental Compliance Reference Guide," implemented the department-wide Inland Bays Recovery Initiative, and administered the state's Coastal Management Program. He began his career in 1976 with the University of California Sea Grant College Program, where he served as an area marine extension agent and coastal resources specialist. He became Director of the University of Delaware's Sea Grant College Program in 1980, and was its Executive Director from 1984–1989. He holds a B.S. degree from the University of New Hampshire and a M.S. degree from Texas A&M University. In their spare time, he and his wife Lynn work hard at maintaining his mother-in-law's farm, where they breed, raise, and train Labrador retrievers.

MAYA K. VAN ROSSUM is the Delaware Riverkeeper and executive director of the Delaware Riverkeeper Network (DRN) since 1996. She is an environmental attorney, strategist, community organizer, facilitator, coalition builder, and manager. As the Delaware Riverkeeper, she serves on a number of the region's committees including the Delaware River Basin Commission's Toxic Advisory Committee, Water Quality Advisory Committee, and its Total Maximum Daily Load (TMDL) Implementation Advisory Committee; the Lower Delaware River Wild and Scenic Management Plan Committee and Advisory Committee; and has served on New Jersey's Stormwater Focus Group. She has been appointed by DEP Secretary McGinty to the Pennsylvania Stormwater BMP manual oversight committee, by Governor Rendell to Pennsylvania's Delaware River Basin Regional Water Resources Committee. Since spring 2002 she has served as an adjunct professor and director of the Environmental Law Clinic at Temple's Beasley School of Law. She served as faculty on the Pennsylvania Bar Institute's 2003 Environmental Law Forum, Water Quantity/Sprawl. She testified before the U.S. House

Committee on Resources concerning wild-and-scenic designation for the Lower Delaware River in 2000. She has a B.S. from La Salle University; a juris doctor degree with certificate in environmental law, cum laude, from Pace University School of Law; and a master's of law degree in corporate finance from Widener University School of Law.

ROBERT WAYLAND retired from the U.S. Environmental Protection Agency in February 2003 where, for twenty-eight years, he served in a variety of key positions, the most recent of which (1991–2003) was director of the Office of Wetlands, Oceans, and Watersheds. In that capacity he exercised national program leadership for wetlands protection, water quality monitoring, watershed management, non–point source pollution control, and ocean programs including the National Estuary Program and dredged material management, and administered an annual budget of over $300 million. His contributions were recognized by Presidents Clinton and Bush, each of whom awarded him the presidential rank of Meritorious Senior Executive; the Wildlife Habitat Council, which awarded him its President's Award; and the Association of State and Interstate Water Pollution Control Administrators, which recognized him with their Elizabeth J. Fellows Partnership Award. Since his retirement, he has consulted with the U.S. Commission on Ocean Policy and serves on the boards of the Southeast Watershed Forum and the Conservation and Preservation Charities of America. He is a graduate of The George Washington University and enjoys sailing and snorkeling.

HEINZ CENTER STAFF

SHEILA D. DAVID is a consultant for The Heinz Center, where she is managing studies for the Sustainable Oceans, Coasts, and Waterways Program. At The Heinz Center she has worked with committees to produce several studies: *The Hidden Costs of Coastal Hazards* (2001), *Evaluation of Erosion Hazards* (2001), *Integrated Management of the Tempisque River Basin, Costa Rica* (2001), *Dam Removal: Science and Decision Making* (2002), *Human Links to Coastal Disasters* (2002), and *Dam Removal Research: Status and Prospects* (2003). Before joining The Heinz Center in 1997, she served for twenty-one years as a senior program officer at the National Research Council's Water Science and Technology Board, where she was study director for some thirty committees that produced reports on topics such as managing coastal erosion, restoration of aquatic ecosystems, protection of groundwater, wetlands characteristics and boundaries, water quality and water reuse, natural resource protection in the Grand Canyon, and sustainable water supplies in the Middle East. She has served as an adviser to and board member of the Association for Women in Science (AWIS) and as editor of *AWIS* magazine. She is also a founder of the National Academy of Sciences' annual program honoring women in science.

JUDY GOSS is a research assistant for The Heinz Center's Sustainable Oceans, Coasts, and Waterways Program, where she has worked on three other studies: *Dam Removal: Science and Decision Making* (2002), *Human Links to Coastal Disasters* (2002), and *Dam Removal Research: Status and Prospects* (2003). She graduated cum laude with a degree in political science from Mary Washington College in 2001. She currently volunteers with the District of Columbia Urban Debate League in Washington, D.C., where she teaches high school students and teachers about policy debate. She is particularly interested in the intersection of gender and political communication and plans to pursue a graduate degree in communication studies.

REFERENCES

American Planning Association. www.planning.org (accessed January 29, 2004).

Brown, J.S., and P. Duguid. 2000. The social life of information. Boston: Harvard Business School Press.

Carson, R. 1951. The sea around us. New York: Oxford University Press.

———. 1962. Silent spring. Boston: Houghton Mifflin.

Clark, H. 1998. Formal knowledge networks. Winnipeg: International Institute for Sustainable Development.

Clarke, K.C. 2002. Getting started with GIS. 4th ed. Upper Saddle River, NJ: Prentice Hall.

Coastal States Organization. About CSO. http://www.sso.org/cso/aboutcso.htm (accessed January 29, 2004).

Cutter, S.L., B.J. Boruff, and W.L. Shirley. 2003. Social vulnerability to environmental hazards. Social Science Quarterly 84(2): 242–261.

Delaware Coastal Program. 2001. Pea Patch Island heronry region special area management plan: Progress report three years of strategy implementation. Dover, DE: The Delaware Department of Natural Resources and Environmental Control.

Dixon, N.M. 2000. Common knowledge. Boston: Harvard University Press.

Evans, N., M.J. Hershman, G.V. Blomberg, and W.B. Lawrence. 1980. Search for predictability: Planning and conflict resolution in Grays Harbor, Washington. WSG 80-05. Seattle: University of Washington Sea Grant.

Federal Geographic Data Committee. http://www.fgdc.gov/ (accessed January 30, 2004).

———. E-Government Geospatial One-Stop. http://www.fgdc.gov/geo-one-stop/ (accessed February 18, 2004).

Florida Department of Environmental Protection. Florida green lodging certificate links. http://www.dep.state.fl.us/waste/categories/recycling/ (accessed December 3, 2003).

Foster, I. 2000. Internet computing and the emerging grid. http://www.nature.com/nature/webmatters/grid/grid.html (accessed January 30, 2004).

Foundations of Success. http://www.fosonline.org/ (accessed on January 29, 2004).

Genetic Programming.org. http://www.genetic-programming.org/ (accessed January 30, 2004).

Good, J.W., and D. Sowers. 1999. Benefits of geographic information systems for state and regional ocean management. Final report to the National Oceanic and Atmospheric Administration Coastal Services Center. Sea Grant Special Report 99-01. Corvallis: Oregon State University.

Goodwin, R.F., and D.J. Canning. 2001. Linking science to shorelines management: Washington's Coastal Planners Group. http://www.psat.wa.gov/Publications/01_proceedings/sessions/poster/j_goodwi.pdf (accessed May 17, 2004).

Green Hotels Association. What are green hotels? http://www.greenhotels.com/whatare.htm (accessed December 3, 2003).

Hardin, G. 1968. The tragedy of the commons. Science 162:1243–1248.

Heinz Center. 2002. Human links to coastal disasters. Washington, D.C.: The H. John Heinz Center for Science, Economics and the Environment.

Imperial, M.T. 1999. Analyzing institutional arrangements for ecosystem-based management: Lessons from the Rhode Island salt ponds SAM Plan. Coastal Management 27:31–56.

Integrated Coastal Management: Sustainability Research Project, 2001–2002. http://www2.mozcom.com/~icm_proj (accessed January 29, 2004).

Internet2. http://www.internet2.edu/ (accessed January 30, 2004.

Jensen, J.R. 1996. Introductory digital image processing: A remote sensing perspective. 2nd ed. Upper Saddle River, NJ: Prentice Hall.

———. 2000. Remote sensing of the environment: An earth resource perspective. 2nd ed. Upper Saddle River, NJ: Prentice Hall.

Kingdon, J.W 1995. Agendas, alternatives and public policies. New York: Little, Brown and Company.

Locally Managed Marine Area Network. http://www.lmmanetwork.org/ (accessed January 29, 2004.

National Estuarine Research Reserve System: Coastal Training Program. http://www.nerrs.noaa.gov/Training/welcome.html (accessed January 30, 2004).

National Office for Integrated and Sustained Ocean Observations: Ocean.US. 2002. http://www.ocean.us/ (accessed January 30, 2004).

National Science Foundation. Fastlane. https://www.fastlane.nsf.gov/fastlane.jsp (accessed January 30, 2004).

National Spatial Data Infrastructure. http://www.fgdc.gov/nsdi/nsdi.html (accessed February 18, 2004).

National States Geographic Information Council. http://www.nsgic.org/ (accessed February 18, 2004).

National Oceanic and Atmospheric Administration Coastal Services Center. 1999a. Coastal resource management customer survey: 1999. http://www. csc.noaa.gov/survey/ (accessed January 30, 2004).

————. 1999b. Ocean Planning Information System. http://www.csc.noaa.gov/ opis/ (accessed March 11, 2004).

————. 2002. Coastal resource management customer survey: 2002. http:// www.csc.noaa.gov/survey/ (accessed January 30, 2004).

National Oceanic and Atmospheric Administration Office of Coastal Zone Management. 1980. Report to the President: Coastal zone management. Washington, DC.

Nickerson, D., and S. Olsen. 2003. Collaborative learning initiatives in integrate coastal management. Kingston: University of Rhode Island, Coastal Resources Center.

Olsen, S., and V. Lee. 1991. A management plan for a coastal ecosystem: Rhode Island's salt pond regions. In Case studies of coastal management: Experience from the United States (pp. 57–69). Kingston: University of Rhode Island, Coastal Resources Center.

Open GIS Consortium. http://www.opengis.org/ (accessed January 30, 2004).

Open Source Initiative. http://www.opensource.org/ (accessed January 30, 2004).

Pew Oceans Commission. 2003. America's living oceans: Charting a course for sea change. Washington, DC: The Pew Oceans Commission.

Rogers, E.M. 1995. Diffusion of innovations. 4th ed. New York: Free Press.

Rose, R. 1993. Lesson-drawing in public policy. Chatham, NJ: Chatham House.

Salus, P.H. 1995. Casting the net: From ARPANET to INTERNET and beyond. Boston: Addison-Wesley.

Schreiber, D.A. 1998. Distance training: How innovative organizations are using technology to maximize learning and meet business objectives. San Francisco: Jossey-Bass.

Seim, H., F. Werner, M. Fletcher, J. Nelson, R. Jahnke, C. Mooers, L. Shay, R. Weisberg, and M. Luther. 2002. SEA-COOS: Southeast Atlantic Coastal Ocean Observing System. Proceedings of the Oceans 2002 Conference, Oct. 29–31, 2002. Biloxi, MS.

Southeast Atlantic Coastal Ocean Observing System. SEA-COOS. http://www. seacoos.org (accessed March 11, 2004).

Stegner, E.W. 1954. Beyond the hundredth meridian: John Wesley Powell and the second opening of the west. New York: Houghton Mifflin.

Stratton Commission. 1969. Our nation and the sea: A plan for action. http:// www.lib.noaa.gov/edocs/stratton/title.html (accessed February 10, 2004).

Sutter, L.A. 2001. Spatial wetland assessment for management and planning (SWAMP). Publication No. 20129-CD. Charleston, SC: NOAA Coastal Services Center.

Sutter, L.A., J.B. Stanfill, D.M. Haupt, C.J. Bruce, and J.E. Wuenscher. 1999. NC-CREWS: North Carolina coastal region evaluation of wetland significance. Raleigh: North Carolina Division of Coastal Management, Department of Environment and Natural Resources.

Sweeting, A.R., and J. Sweeting. 2003. A practical guide to good practice: Managing environmental and social issues in the accommodations sector. May be accessed from http://www.celb.org/xp/CELB/publications-resources/ (accessed May 17, 2004).

U.S. Commission on Ocean Policy. Preliminary report of the U.S. Commission on Ocean Policy. Governor's Draft. April 2004. Washington, DC: U.S. Government.

U.S. General Accounting Office. 2003. Information technology training: Practices of leading private-sector companies. GAO-03-390. www.gao.gov/cgi-bin/getrpt?GAO-03-390/ (accessed February 18, 2004).

U.S. Geological Survey. The National Map. http://nationalmap.usgs.gov/ (accessed February 18, 2004).

Waterkeeper Alliance. Waterkeeper alliance is . . . http://keeper.org (accessed February 9, 2004).

Wenger, E. 1999. Communities of practice. Cambridge: Cambridge University Press.

Wenger, E., R. McDermott, and W. Snyder. 2002. Cultivating communities of practice. Boston: Harvard Business School Press.